Abnormal Chromosomes

Abnormal Chromosomes

The Past, Present, and Future of Cancer Cytogenetics

Sverre Heim

The Radium Hospital and University of Oslo, Norway

Felix Mitelman

University Hospital and University of Lund, Sweden

WILEY Blackwell

This edition first published 2022

Registered Offices
John Wiley & Sons, Inc., 111 River Street, Hoboken, NJ 07030, USA
John Wiley & Sons Ltd, The Atrium, Southern Gate, Chichester, West Sussex, PO19 8SQ, UK

Editorial Office
9600 Garsington Road, Oxford, OX4 2DQ, UK

For details of our global editorial offices, customer services, and more information about Wiley products visit us at www.wiley.com.

Wiley also publishes its books in a variety of electronic formats and by print-on-demand. Some content that appears in standard print versions of this book may not be available in other formats.

Library of Congress Cataloging-in-Publication Data applied for:

ISBN 9781119651987 (hardback)

Cover Design: Wiley
Cover Image: © koya979/Shutterstock

Set in 10.5/13pt STIXTwoText by Straive, Chennai, India
Printed and bound by CPI Group (UK) Ltd, Croydon, CR0 4YY

C9781119651987_240122

Contents

Preface

In 1956, when we were still very young, the chromosome number of humans changed from 48 to 46. Not correct, some may counter, no such thing happened, only a previous misconception was rectified when better cytogenetic analyses were brought to bear. Of course they are right, those who maintain that the change was one of knowledge, not biology itself; no example of saltational evolution was witnessed, no actual sudden leap in the ever-lasting development of our species took place. And yet one wonders: How many other, vaguely similar mistakes form part of our current picture of the human genome during health and disease in spite of the tremendous strides forward taken by science during the last century?

Human cytogenetics came of age during the latter part of the twentieth century with regard to both investigations of the constitutional human karyotype and the acquired chromosomal abnormalities that characterize neoplastic cells. The two authors of this book have worked scientifically as well as clinically within cancer cytogenetics during much of this period. Our own investigations and findings were made public in numerous articles. In four books, all entitled *Cancer Cytogenetics* and published between 1987 and 2015, we tried to overview the field. With each year and publication, the *status praesens* of cancer cytogenetics changed.

The purpose of the present book is different. Instead of concentrating on presenting in considerable detail the current body of cytogenetic knowledge as applied to neoplastic conditions, we also want to look backward as well as forward, to the past as well as the future. How did the study of chromosome aberrations in neoplastic cells develop? What modes of thinking and technological advances paved the way for the cytogenetic discoveries that were seminal in forming today's understanding of how cancer develops, how it should be diagnosed, the prognostic impact of various genomic aberration patterns, and, not least, which treatment should be chosen?

We try to be fair in our coverage of what happened, but of course the story we tell is colored by what *we* have found interesting over the years, what *we* see as important. Inevitably, others may have viewed the same events in a different light and would, hence, weigh the available information differently.

This is always the case when stories, even scientific ones, are told, for science is by no means exempt from the various personal imprints and biases that characterize all written accounts.

On the other hand, we have no particular ax to grind; at least, we are not ourselves aware of any such bias be it scientific or otherwise. Whether those who read will agree with us or not is a different matter. Readers and writers alike have a tendency to tinge any text according to their own experience and underlying preferences, a fact better acknowledged up front.

When covering the present role of cancer cytogenetics (the second part of the book) and, even more difficult, trying to extrapolate from what is into what will be (Part Three), perspective becomes even more crucial. We choose to focus on three areas in particular: How cancer cytogenetic analyses contribute to a better understanding of tumorigenesis, how they are of value clinically, and how they meet, interact with, and cross-fertilize other investigative technologies, in particular those of molecular genetics.

The first two points are uncontroversial, for it is widely accepted that cancer is a genetic disease at the cellular level, and that acquired genomic aberrations lie at the heart of neoplastic transformation. Many of the said aberrations were first detected by chromosome analysis whereafter molecular investigations detected their submicroscopic details and consequences. Likewise, several of the new therapeutic techniques that form part of personalized medicine target cancer-specific genomic alterations, something that makes them specific far beyond what could be obtained with anticancer drugs of the past. The search for new such changes goes on, as do the efforts to develop suitable drugs. Also for this reason, future classifications of cancer are likely to be much more pathogenesis-based than is the case today.

More uncertain is the future role of a basically microscopy-dependent technique such as cancer cytogenetics, for is it not so that the modern synthesis between molecular genetics and computerized technologies is making superfluous anything so crude as to involve actual human eye inspection of subcellular structures? Maybe, but maybe not; it is not within our abilities to foresee the future, nor can anybody else perform such a feat. Some observations of what has already come to pass may nevertheless help us assess the various claims and predictions that are tossed so freely about, and for which their originators are rarely, if ever, held to account once they are shown to be incorrect.

Already in the 1980s, predictions were abundant that new techniques would soon – within the span of only a few years, some said – make cytogenetics obsolete. This did not happen, although admittedly many important new questions about nature could be answered using the molecular techniques introduced at that time. There is nothing strange in such a change; in fact, cytogenetics, molecular genetics, and clinical cancer medicine benefited immensely and reciprocally from the combined use of the new and the old. Not always was the technological mix optimally calibrated, however. Both undue reluctance to introduce novel methods and the opposite, an overeager attitude toward all that is new and sparkling, can be harmful to well-balanced quests for better medical-biological understanding. We discuss some such instances in the latter parts of this book.

The necessary complementarity of morphological and functional investigations of biological systems, at different levels of resolution or organization, in our opinion remains central to any balanced approach to the study of complex diseases such as cancer, regardless of which techniques are more fashionable at any given time. Seeing will never become obsolete to the researcher who wants to understand; hence, cancer cytogenetics will never die. Whether its future becomes more of a legacy or if it continues to play an integral part in state-of-the-art analyses is unknown. It could depend on whether new understanding emerges as to how topological genome features regulate the higher-order orchestration of gene activities, possibly combined with new ways of studying chromosomes at the time when they perform their normal work, i.e., during interphase. At present, such developments are largely conjectural. Only fools think they can foresee the future, and we would rather not come across as fools.

Regardless of the extent to which microscopic searches for the chromosome abnormalities of transformed cells will remain important or not, the cytogenetic history of human oncology deserves to be told. We hope you will enjoy our version of it. It would not surprise us if later stories on the same topic, told by our successors, were to paint a different picture. What we today glimpse through futuristic binoculars is in all likelihood going to look very different in hindsight. The normal chromosome number of humans may not be likely to undergo another change, but many other "truths" will. Progress never stops, not even when the subject matter under study remains unaltered.

An informatory note should be added about the "literature" listed at the end of each chapter. Unlike what is customary in articles and books of a scientific nature, we have not striven to back up all statements, be they controversial or not, with references, although some are given. By way of compensation for the incomplete referencing, but also to broaden the scope of our coverage generally, most chapters contain suggestions of review articles and books that interested readers may find valuable. Our latest multiauthor textbook *Cancer Cytogenetics: Chromosomal and Molecular Genetic Aberrations of Tumor Cells* (4th edition) more likely than not contains all the missing references to primary literature, at least up to 2015. Also the regularly updated Mitelman Database of Chromosome Aberrations and Gene Fusions in Cancer (https://mitelmandatabase.isb-cgc.org) contains exact information on abnormal karyotypes and their molecular consequences, as well as complete reference lists.

Oslo and Lund, July 2021

Sverre Heim
Felix Mitelman

PART I

PAST

CHAPTER 1

Understanding Disease: Ancient Theories

For a number of good reasons, man's attitude toward disease and death constitutes an important part of his attitude toward life itself. It cannot be otherwise, for no other phenomena are more existentially threatening, more essential to our being and nonbeing, than health problems. The development of what we today call medicine has always been central to our best thinking about our best interests: What is the essence of life and death, what causes different diseases, how do they develop, how are they best combatted and even, in the more fortunate cases, how can the ill be healed?

As far as we know, the first attempts at answering these and similar questions invoked magic. Gods or demons or other supernatural beings intervened for whatever good reason they had and created problems for us, or solved them, of their own volition and whenever they saw fit. Our own abilities to influence the resulting processes were thus restricted to attempts at pleasing the more powerful amongst them – bringing sacrifices, offering prayers and what not – which people duly did according to the habits of their society.

A problem with this principle of disease causation, however, is the many-to-many relationships it entails (many demons or gods, many ailments), and which inevitably preclude any possibility of reliably predicting things or results, be they favorable or the opposite, of both diseases and attempts to cure them. Everything was haphazard, dependent on the will and whims of supernatural beings and our ability to make them happy. Such

Abnormal Chromosomes: The Past, Present, and Future of Cancer Cytogenetics.
Sverre Heim and Felix Mitelman.

a scheme of things is simply not compatible with a rational approach to how the universe is put together and how health and disease have their respective places in a natural world.

We experience chaos around us, a multitude of phenomena that appear to be without connection or meaning. The ancient Greeks were the first to try to sort things out in a manner we today would call rational: Attempts to create order out of disorder, or cosmos out of chaos, were the predominant leitmotif of all their strivings and remain the underlying foundation also of later scientific efforts.

Natural philosophers set out to find the putative one substance out of which everything was made. Some said water, others air and so on; even highly abstract and modern-sounding concepts were brought to bear, like Anaximander's infinite (*to apeiron*) in the fifth century BCE. As with scientists of later times and their explanatory attempts, there was never a shortage of alternatives or disagreement among thinking people's accounts of reality, not even at the very beginning. Compromises were reached between those who sought a single material first cause and the phenomenologically inclined who recognized only the totality of things or events; Empedocles, who lived at about the same time as Anaximander, pointed to earth, wind, fire, and water as the four elements from which everything was made.

A similar *tetrad* gained prominence in thinking about disease causation during antiquity, namely that a disturbance of the balance between the body's four basic fluids – black and yellow bile, blood, and phlegm – caused all our health problems. Later on, a fusion occurred between this classification and Empedocles' elements, and different prevailing temperaments – resulting in people being thought to be predominantly *choleric*, *melancholic*, *phlegmatic* or *sanguinic* – were singled out corresponding to which imbalance among them existed within the body. For a long while, all of 2000 years, and for reasons that are anything but clear, the number four thus seemed to enjoy preferential treatment in the pathogenetic thinking of learned men of medicine, though competing theories drawing on the alleged existence of more or fewer humors or elements certainly existed.

The father of medicine, Hippocrates (around 460–370 BCE), is the one to whom these *humoral theories of disease* are usually ascribed. Whatever their validity and Hippocrates' actual historical role, to him belongs the traditional honor of having introduced rationality into medical thinking. No disease was caused by supernatural influences, spells and their like; even the sacred disease, what we today call epilepsy, was but a natural disorder in Hippocrates' opinion. In this sense, he was truly the conceptual father of scientific medicine from which cancer cytogenetics flows as but one of numerous present-day offsprings.

While ancient thinking about cancer and other diseases characterized clinically by *phtisis* (wasting) was of necessity limited by the total absence of what we today would deem useful investigative tools, some principles of lasting importance can nevertheless be extracted from the very brief history drawn up above. Diseases, cancer included, are rational phenomena that can, at least in principle, be understood. Their causation is difficult, but the reduction of causes to as few as possible should be attempted (Ockham's "razor principle" from the high Middle Ages comes to mind: "Do not multiply causes unnecessarily!"). Finally, understanding of disease processes is a prerequisite for successful treatment; only in the most fortuitous of cases can one count on being able to achieve a cure if nothing of essence is known about the disease one is faced with. All these themes are going to be visited in the chapters to come.

FURTHER READING

Lane Fox, R. (2020). *The Invention of Medicine: From Homer to Hippocrates.* London: Penguin.

Porter, R. (2006). *The Cambridge History of Medicine.* Cambridge: Cambridge University Press.

Wise Bauer, S. (2015). *The Story of Western Science: From the Writings of Aristotle to the Big Bang Theory.* New York: WW Norton.

CHAPTER 2

The Advent of Cellular Pathology

Good thinking alone is rarely enough to gain reliable insights into the workings of nature, neither in general nor when it comes to what causes illness and death; also relevant experimental data are needed. Brave and curious men for centuries cut up and examined corpses of the recently deceased to see with their own eyes (which is the exact meaning of the word "autopsy") what characterizes victims of various diseases.

This activity was looked down upon, even viewed as a crime in many societies and over long periods of time, and yet our present knowledge of both normal and pathological anatomy owes a large debt of gratitude to these courageous pioneers. Perhaps most prominent among them was Andreas Vesalius (1514–1564) who is sometimes referred to as the founder of human anatomy. His *De humani corporis fabrica* ("On the structure of the human body") represented a major step in the establishment of scientific medicine and long remained one of the most influential books on human anatomy (Figure 2.1). It was exceptional in building exclusively on information gained from examination of human corpses, not dissection of monkeys or other animals as had previously often been the case. It is a bemusing coincidence that Vesalius' revolutionary treatise was published in 1543, the same year that Copernicus' *De revolutionibus coelestium orbitum* ("On the revolutions of the celestial spheres") came out, practically on the author's death bed. The mathematician and astronomer – who was also a good Catholic canon – thus saw to it that no one would be able to prosecute him for the blasphemy

Abnormal Chromosomes: The Past, Present, and Future of Cancer Cytogenetics.
Sverre Heim and Felix Mitelman.

FIGURE 2.1 Woodcut anatomical illustration of the muscular system from Vesalius' 1543 masterpiece *De humani corporis fabrica*. Source: Wikipedia/Public Domain.

of ascribing to the sun, not the earth, prime position within the universe. Indeed, those were heady times, with Renaissance man taking long strides in several directions toward a deeper understanding of reality. Some of the directions were wholly novel while others represented rediscoveries – or rebirths – of thoughts first voiced by the ancients.

Vesalius and others described – meticulously, honestly, and in considerable detail – what we today call macroscopic human anatomy. Scientific knowledge about pathogenetic (*pathogenesis* means how a disease develops whereas *etiology* deals with why it occurs) changes taking place beyond the resolution level of human eyesight had to await the introduction of technological novelties that augment our innate senses, above all the art of how to grind lenses and arrange them to see what had hitherto been invisible.

Whereas simple magnifying glasses, for example water-filled spheres, had been used occasionally since antiquity and eyeglasses with primitive lenses since the thirteenth century, the first certain examples of the use of compound microscopes, combining an objective lens near the specimen with an eyepiece to view a real image, date from the first decades of the seventeenth century. Some of the names associated with these initial telescopic and microscopic studies are among the absolute giants of modern science – Galileo Galilei's discovery of Jupiter's moons (amongst other things) and subsequent controversy with the church over the validity of Copernicus' heliocentric system come first to mind – whereas others are less well known today. Among the latter are particularly many Dutchmen, for Amsterdam was the center of lens-grinding technologies as well as the assembly of instruments made from them. The industry was not without its dangers, and not only because of whatever misgivings the powers that be might have about the discoveries scientists made using state-of-the-art equipment put together in Amsterdam's workshops, but also for the manual workers. The famous philosopher Baruch de Spinoza seems to have been one such victim. In order to pursue his intellectual interests without being dependent on the rich and opinionated, he had taken up the craft of lens grinding. In 1677, at the young age of 44, he died from a lung disease that may have been caused by inhalation of glass dust, though tuberculosis remains another possibility.

Examination of biological specimens by many investigators in the mid-seventeenth century (Anton van Leeuwenhoek's contributions seem to have been particularly valuable) using increasingly refined microscopes – first with single lenses, later in combinations – eventually led to the identification of cells, a discovery usually credited to Robert Hooke in 1665. The name derives from Latin *cella* meaning "a small room," something akin to the ones monks lived in. However, due to the still insufficient magnification obtained, microscopists could not yet see any internal components of the cells they studied. Thus, nobody at the time had any clear conception of the cells' structure or function, let alone one backed up by what we today would call solid scientific evidence.

This did not prevent some researchers from holding strong views based on what they claimed to see. The story about the *homunculus* (Figure 2.2) – and *animalcules* in other species – is a case in point. Nicolas Hartsoeker was a lens grinder who had studied optics under van Leeuwenhoek and become an expert microscope builder. Toward the end of the seventeenth century, he conducted the first known microscopic studies of human semen. Hartsoeker "saw" within the sperm cells' head a tiny person, one *homunculus* per head, who he assumed was destined to grow into a full human being after reception and subsequent nurturing by the fertile female soil. This erroneous observation or interpretation seemingly confirmed the spermist theory of conception which held sway for centuries (Paracelsus appears to have been the first medical authority to have stipulated the existence of homunculi, in *De rerum natura* from 1537). The above rendering of the story may not be entirely precise, however, as pointed out in the caption to Figure 2.2, so perhaps Hartsoeker deserves to be at least partly

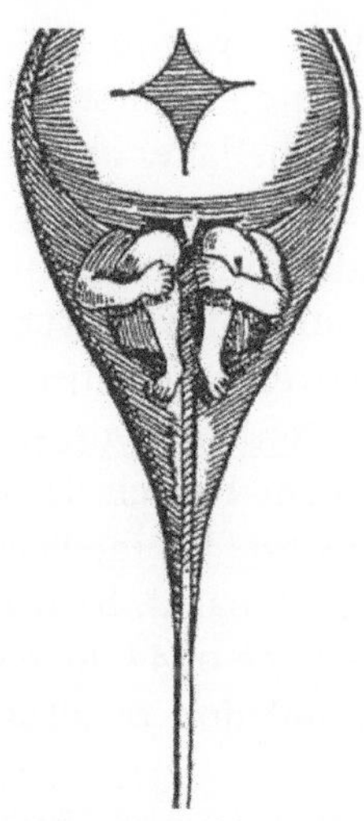

FIGURE 2.2 A tiny, preformed human inside a sperm – a *homunculus* – drawn in 1694 by the Dutch microscopist Nicolas Hartsoeker. This figure has been reproduced countless times, usually with the caption stating that it represents the homunculus Hartsoeker saw, or thought he saw, under the microscope. Yet it seems that Hartsoeker only said that "perhaps" we would see this if it had been possible to see through the "skin" that surrounds the sperm head, and "if we had the tools." The exact story behind the drawing will probably never be known. Source: Wikimedia Commons/Public Domain.

exonerated. At any rate, the long-lasting homunculus *intermezzo* illustrates that strange ideas about (im)balances between the two sexes clearly are not peculiar to modern times.

The quality of microscopes did not change significantly from the period of Hooke and Leeuwenhoek until the 1800s, although incremental improvements in the microscopists' picture of what cells and tissues look like – the two fields became known as *cytology* and *histology*, respectively – of course occurred. Worthy of mention in this context was the discovery by Karl Rudolphi and J.H.F. Link that cells have independent, not shared, walls as had hitherto been assumed. For this, in 1804 awards were bestowed upon them by the Royal Society of Science, Göttingen, Germany, for having "solved the problem of the nature of cells," no less. The same year, Franz Bauer provided compelling evidence for the existence of a cell nucleus.

Out of all these studies grew the understanding that cells are the fundamental elements of life itself. This so-called cell theory was eventually formulated in 1839 by Matthias Schleiden (a botanist) and Theodor Schwann (a physiologist): All living organisms, be they plants or animals, are composed of one or more cells which thus constitute the basic units of vital structures. Today we sometimes recognize also noncellular entities such as viruses as forms of life, but otherwise the cell theory holds true.

Schleiden, like many others before him, originally thought that free cell formation occurred through crystallization, but that hypothesis was refuted in the 1850s by several investigators who instead found that cells themselves give rise to new cells, by division or binary fission. Shortly afterwards, the German pathologist Rudolf Virchow (1821–1902) formulated this new insight into one of cell theory's most central tenets: *Omnis cellula e cellula* (everything cellular stems from cells).

It is worthy of note that the use of microscopes was not universally embraced by biologists, just as in Galilei's time there were some astronomers who did not consider telescopes reliable. The cell theory came into being at a time when histology was still dominated by the teachings of the French anatomist Bichat who thoroughly distrusted the use of microscopes and, consequently, whatever they helped examiners to see. Based on gross investigations alone, Bichat described no fewer than 21 different types of animal tissues. Of necessity, that purely macroscopy-based classification was not reconcilable with the wave of new data coming from microscopic examinations.

Rudolf Virchow (Figure 2.3) assumed a leading role in the revolution in pathological understanding that the use of high-quality microscopes produced by Zeiss, and later also other companies, enabled. He was not only a medical doctor, but also an anthropologist and a politician who campaigned vigorously for social reforms and even served in the German Reichstag for more than 10 years. In the latter role, he allegedly angered Otto von Bismarck so much that the Iron Chancellor challenged him to a duel. When Virchow suggested as weapons two pork sausages, one of which was infected with *Trichinella*, to be chosen between and eaten, Bismarck reportedly refused to participate. Another version of the story holds that Virchow declined because he considered dueling an uncivilized and irrational way to solve a conflict.

Virchow's most lasting legacy is as the father of cellular pathology, however. He made innumerable contributions to the field, including a description of how blood clots in the legs could dislodge and become emboli that later became stuck in the lungs, and even coined the very terms "thrombus" and "embolus." But in the context of this book, we are more concerned with his understanding of cancer as a cellular disease. Malignant tumors occur because malignant cells divide, Virchow maintained, in an uncontrolled manner, causing destruction of surrounding tissue, even spreading as emboli through blood and lymph vessels to set up metastases in

FIGURE 2.3 Rudolf Virchow in his office, 1901. Source: Mary Evans Library/Adobe Stock.

distant organs. He was also the first person to recognize leukemia. Seeking a name for this condition, Virchow first and logically settled on “weisses Blut.” In 1847, he changed the name to the more academic-sounding “Leukämie” (from *leukos*, the Greek word for “white”).

Exactly how the various neoplastic processes (*neoplasia* means “new growth” in Greek) occur, beyond the fact that it has to do with too many cells accumulating, not with fluid imbalances or anything similar, remained a moot point throughout the nineteenth century though more or less esoteric theories certainly abounded. Both cancer and benign neoplasms became established as examples of disease processes solidly placed within cellular pathology, although the path to more certain and detailed pathogenetic knowledge could only be trod by researchers in the generation(s) after Virchow. The proper studies to address these questions would look specifically at what takes place within the cells, even within their nuclei.

FURTHER READING

Gest, H. (2004). The discovery of microorganisms by Robert Hooke and Antoni van Leeuwenhoek, fellows of the Royal Society. *Notes Rec. R. Soc. Lond.* 58: 187–201.

Ghosh, S.K. (2015). Human cadaveric dissection: a historical account from ancient Greece to the modern era. *Anat. Cell Biol.* 48: 153–169.

Gribbin, J. (2004). *The Scientists: A History of Science Told through the Lives of its Greatest Inventors*. New York: Random House.

Mazarello, P. (1999). A unifying concept: the history of cell theory. *Nat. Cell Biol.* 1: 13–15.

Nadler, S. (2000). Baruch Spinoza. Heretic, lens grinder. *Arch. Ophthalmol.* 118: 1425–1427.

Ribatti, D. (2018). An historical note on the cell theory. *Exp. Cell Res.* 364: 1–4.

Scarani, P. (2003). Rudolf Virchow (1821–1902). *Virchows Arch.* 442: 95–98.

Turner, W. (1890). The cell theory, past and present. *J. Anat. Physiol.* 24: 253–287.

CHAPTER 3

The Colored Bodies of Cell Nuclei: Chromosomes and Heredity

While Virchow and his colleagues established cellular pathology as the dominant mode of thinking about disease processes, other lines of research led to no less significant conceptual breakthroughs in our understanding of how biological systems undergo incremental change over the long haul despite the amazing stability maintained when moving from one generation to the next. The publication of Charles Darwin's "On the Origin of Species by Means of Natural Selection, or the Preservation of Favoured Races in the Struggle for Life" (1859) stands out as particularly important; Darwinism was to have an immense impact on society philosophically, scientifically, and politically. The central principles of this *theory of evolution* – not only are populations of living organisms, for example within a given species, mutable but those individuals or subgroups of organisms that somehow acquire useful heritable features inevitably tend to take over ("survival of the fittest") – would later be accepted to apply also at the level of cells. Phenotypic selection for genotypically determined abilities in neoplastic cell populations (the *genotype* is the genetic constitution of an individual whereas the *phenotype* is the sum of manifestations that same genotype gives rise to while interacting with environmental factors), including how different environments over time may select for different, often genetically complex subpopulations, will be a recurring theme of this book.

Abnormal Chromosomes: The Past, Present, and Future of Cancer Cytogenetics.
Sverre Heim and Felix Mitelman.

In the mid-1860s, two publications came out that in the field of genetics would assume an almost equally paradigmatic role as Darwin's work had within general biology. Francis Galton (who happened to be Charles Darwin's half-cousin) studied the heritability of various forms of mental excellence in prominent English families; including his own, we might add. In "Hereditary Talent and Character" (two articles from 1865), he presented a biometric measurement of complex phenotypic differences and, perhaps even more importantly, offered methods as to how genetic conclusions could be drawn from the data thus acquired. Galton's work became a scientific starting point for quantitative genetics, the mathematics-based study of polygenic traits or, more precisely, of multifactorial (when both multiple genes *and* the environment influence the trait under scrutiny) characteristics.

At the same time, Gregor Johann Mendel, an Augustinian friar and abbot of St Thomas' abbey in Brno (Brünn), Moravia, then part of the Austrian empire, diligently pursued his scientific interests in botany, more specifically how different traits of peas are distributed after various crossings of pea plants. His findings were published in 1865 ("Versuche über Pflanzenhybriden") in an obscure journal, *Verhandlungen des naturforschenden Vereines in Brünn*, whereupon they were promptly forgotten (if noticed at all) by the scientific community until rediscovered simultaneously in 1900 by Carl Correns, Hugo de Vries, and Erich von Tschermak.

It may be worthy of note that quite a few subsequent historians of science – on various grounds, we might add, but in numerous articles as well as books – have questioned the relative contributions of Correns, de Vries, and von Tschermak to the rediscovery of Mendelian genetics. We are not in a position to assess the validity of their, sometimes contradictory, criticism. The sequence of events related above is the classical one that, barring new evidence to the contrary, will have to suffice. All in all, the controversy over who rediscovered Mendel's work and when is but one of many examples of how even historical events are complex and ridden with strife about priority.

Whereas Galton studied obviously interesting traits like human ability to do mathematics and play music, finding only complex relationships hinting at the hereditary basis of such faculties, Mendel had the genius or serendipity to study something thoroughly uninteresting (at least to nonbotanists!), namely seven phenotypic characteristics of peas, including whether they were smooth or wrinkled. To make a long and often-told story short: Based on his observations, the counting of peas, and a piece of truly excellent synthetic thinking, he formulated laws for how single-gene or Mendelian genetics works. In the process, he also coined essential terms such as *recessive* and *dominant* for hereditary traits and explained how the elusive "factors of heritability" (the name *gene* for these factors came into use only in 1909, introduced by the Danish botanist Wilhelm Johannsen) must behave during cell division to achieve the observed regularities.

It was thus only when attempts were made in the beginning of the twentieth century to obtain satisfactory models of *how* discontinuous inheritance operates (continuous or blended inheritance had been well accounted for by Galton's quantitative genetics) that Mendel's contributions were finally brought to the fore of scientific attention and recognized for their inherent explanatory power. At last, Mendelian thinking now *did* meet with widespread acceptance, especially through the work of William Bateson who is also famous for having introduced the term "*genetics*."

Since we mentioned above that controversies have raged around the "true history" of how Mendelian genetics was rediscovered, it may be prudent to mention, at least in passing, that doubts have also been voiced about the statistical validity of Mendel's original pea experiments. Thus, the prominent population geneticist R.A. Fisher in 1936 opined that Mendel's counts were "much too good to be true." Either the founding father of modern single-gene genetics had somehow been the victim of severe observation bias or, the most likely explanation, his data had been deliberately "improved," possibly by an unknown assistant. Fisher's accusations did not receive much attention at first, but from 1964 on, around the centenary of "Versuche über Pflanzenhybriden," scholars began to discuss whether the fielded criticism really meant that Mendel's data had been falsified. The majority opinion seems to hold that although grounds exist to assume that a certain unconscious bias may have influenced Mendel's experimentation, suspicions of deliberate falsification remain unproven.

Be all that as it may: Today we know that Mendel's laws are more than just another inspired theory; they plain and simple reflect how things are with only very few exceptions that need not worry us in the present context, and which could not possibly have been foreseen at his time. From our perspective rooted in nonstatistical, laboratory research, we see as essential not exactly *how* Mendel arrived at his results statistically, but that they led to insights into essential biology at a depth since unsurpassed. The man was *right*, and that is the most important thing. He was as great a hero and as true a genius in the history of biology as there ever was. Mendel's own reaction when realizing that his contemporaries did not pay much attention to his theory of heredity was allegedly the stoical "my time will come." In that prediction, too, he was amazingly accurate.

Could single-gene genetics find a proper place within the cellular theory of life, thus uniting the two seemingly separate branches of biology? It turned out that indeed it could, though quite some time would pass and a lot of experimentation had to be performed before the key correlations between cellular form and behavior, on the one hand, and on the other, the transfer of hereditary characters from one generation to the next, were detected.

It seems that the German botanist Hugo von Mohl was the first (1835) to describe the splitting of a mother cell into two daughter cells, what we now call *cell division*. He studied green algae. Significant progress in the microscopic study of nuclear anatomy and cell division came when, in the 1870s, the anatomist Walther Flemming began to stain nuclei with basophilic aniline dyes. The cerulean-shaded intranucleic material he saw (similar observations had apparently been made by Edouard van Beneden and Eduard Strasburger) was named *chromatin* (after the Greek word for color). These thread-like structures, which were also called "chromatic elements," "nuclear segments" or "karyosomes" by different researchers, were finally (1888) given the name *chromosomes* (hence the "colored bodies" of the chapter heading) by Wilhelm von Waldeyer-Hartz.

Investigating cell division, Flemming noted that chromosomes were distributed to daughter nuclei (Figure 3.1) in a process he dubbed *mitosis* (Greek for thread). His studies were published (1882) in the book *Zellsubstanz, Kern und Zelltheilung* (Cell Substance, Nucleus and Cell Division) in which he hypothesized that all nuclei had to originate from preceding nuclei. He therefore rephrased Virchow's "Omnis cellula e cellula" into "Omnis nucleus e nucleo," putting additional emphasis on the central role of the nucleus, and thus hereditary material, in all of biology. Flemming fully deserves to be considered one of the founders of *cytogenetics*, a branch of cytology (cell science) dealing with the study of chromosomes.

As is typical of fundamental cell biology research, chemical studies proceeded in parallel with anatomical, microscopic investigations. Particular credit should in this context be given to the Swiss physician and biologist Johannes Friedrich Miescher who studied the chemistry of cell nuclei. Around 1870, Miescher isolated from dead polymorphonuclear leukocytes present in pus-soaked bandages a precipitate he called *nuclein*. Analysis of the substance revealed that it contained a lot of phosphorus and nitrogen but, surprisingly at the time, no sulfur. It was subsequently shown that nuclein in the main consisted of deoxyribonucleic acid or DNA, the molecule that much later, in 1944, was shown in the Avery–MacLeod–McCarty experiment to be the very essence of heredity.

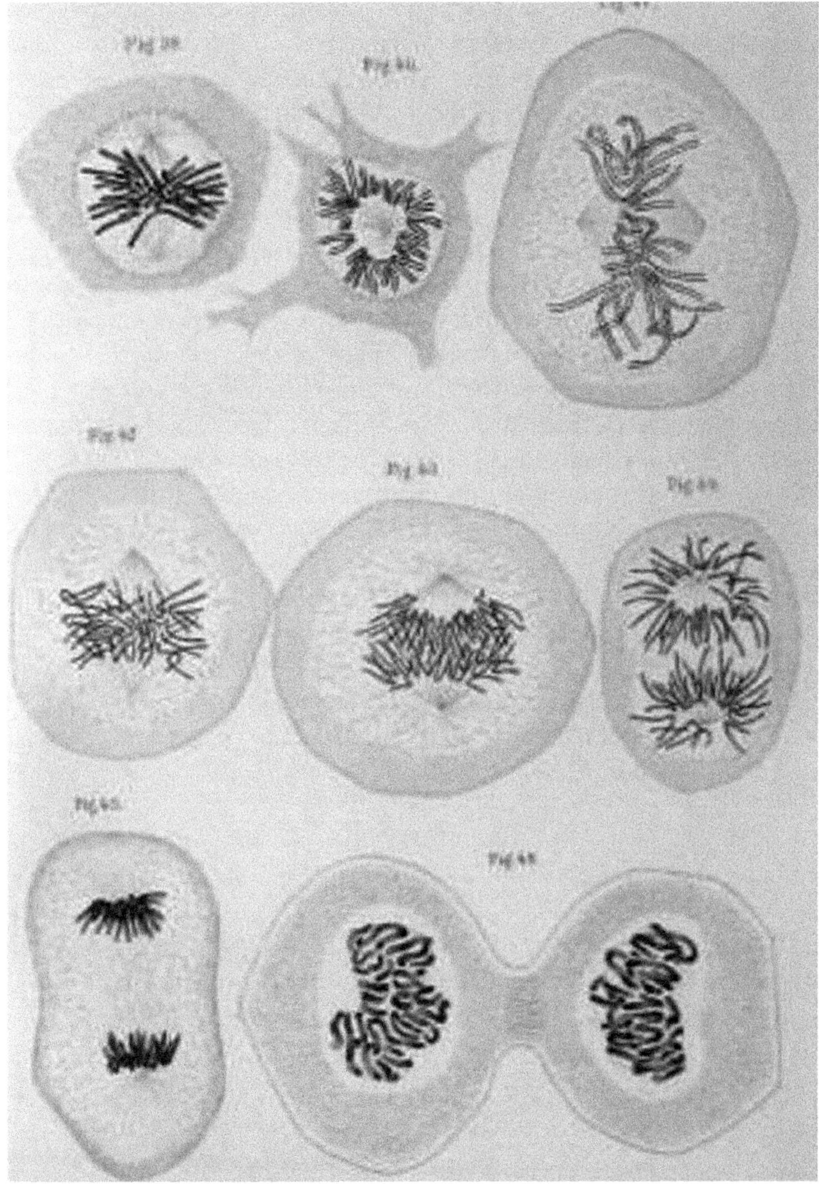

FIGURE 3.1 The earliest known drawings of human chromosomes based on observations in dividing cells published by W. Flemming in his book *Zellsubstanz, Kern und Zelltheilung* (Vogel, Leipzig, 1882).

Another important cytologist and cell physiologist of the era was Hertwig who, studying sea urchins, in 1876 proved that fertilization occurs by fusion of sperm and egg cells; the former cell penetrates and is absorbed into the latter. Hertwig then, in 1885, claimed that nuclein was the substance responsible not only for fertilization but also for the transmission of hereditary characteristics.

That the hereditary material is indeed located within or on the aforementioned colored intranucleic bodies constitutes the central tenet of the *chromosome theory of inheritance* advocated by several researchers in the first decade of the twentieth century. Work by, in particular, Walter Sutton and Theodor Boveri linked heredity, or the new science of genetics, with cytological observations in a manner that provided Mendel's laws with a much needed anatomical backbone. The thread-like chromosomes seen by Flemming and others were identified as the nucleic structures on which Mendel's paired factors, now called genes, were sitting in a linear manner at specific sites called *loci*. How chromosomes were observed to enter daughter cells during *mitosis* as well as during germ cell division, *meiosis*, now made several features of genetics understandable; chromosomes and their genes (the latter were still putative inasmuch as they did not yet have any known chemical or anatomical correlate) were "such stuff as genetics are made of." It did not take long before all this new knowledge about heredity was used to help explain the pathogenesis of various diseases, amongst them cancer, a process in which one of the originators of the *Boveri-Sutton chromosome theory of inheritance* would play a major role.

FURTHER READING

Cremer, T. and Cremer, C. (1988). Centennial of Wilhelm Waldeyer's introduction of the term "chromosome" in 1888. *Cytogenet. Cell Genet.* 48: 66–67.

Forrest, D.W. (1974). *Francis Galton: The Life and Work of a Victorian Genius.* New York: Taplinger Publishing Co., Inc.

Franklin, A., Edwards, A.W.F., Fairbanks, D.J. et al. (2008). *Ending the Mendel–Fisher Controversy*. Pittsburgh: University of Pittsburgh Press.

Gribbin, J. (2004). *The Scientists: A History of Science Told through the Lives of its Greatest Inventors*. New York: Random House.

Gustafsson, Å. (1969). The life of Gregor Johann Mendel – tragic or not? *Hereditas* 62: 239–258.

Mukherjee, S. (2017). *The Gene: An Intimate History*. New York: Scribner.

Paweletz, N. (2001). Walther Flemming: pioneer of mitosis research. *Nat. Rev. Mol. Cell Biol.* 2: 72–75.

Stubbe, H. (1972). *History of Genetics, From Prehistoric Times to the Rediscovery of Mendel's Laws.* Cambridge: MIT Press.

Sturtevant, A.H. (2001). *A History of Genetics.* Long Island: Cold Spring Harbor Laboratory Press.

CHAPTER 4

Boveri and the Somatic Mutation Theory of Cancer

David Hansemann (1858–1920), one of Virchow's assistants at the Department of Pathology in Berlin, the international center of cellular pathology toward the end of the nineteenth century, would be the first to report (1890) on the nuclear and mitotic irregularities of cancer cells. Studying sections from 13 different carcinomas (carcinomas are malignant tumors displaying epithelial differentiation whereas sarcomas are their malignant connective tissue counterparts), he observed frequent abnormal cell divisions which sometimes led to increased, sometimes decreased chromatin content in the daughter cells (Figure 4.1). His observations and conclusion, that these phenomena were not only diagnostically important but central to the very development of malignancies, mark the beginning of cancer cytogenetics.

The next conceptual leap forward in the story of how cancer cytogenetics evolved would come from an unlikely source, a person who had no experience with cancer pathology and indeed was not even a medical doctor. The German zoologist Theodor Boveri (1862–1915) instead studied exotic organisms, from a medical point of view, such as sea urchins and the eggs of *Ascaris*, for which he traveled to the Stazione Zoologica in Naples, Italy. He noticed that the cells he examined depended on having a correct chromosome complement for proper embryonic development to take place. He also, while he was still in his 20s, discovered the nuclear *centrosome* and identified it as

Abnormal Chromosomes: The Past, Present, and Future of Cancer Cytogenetics.
Sverre Heim and Felix Mitelman.

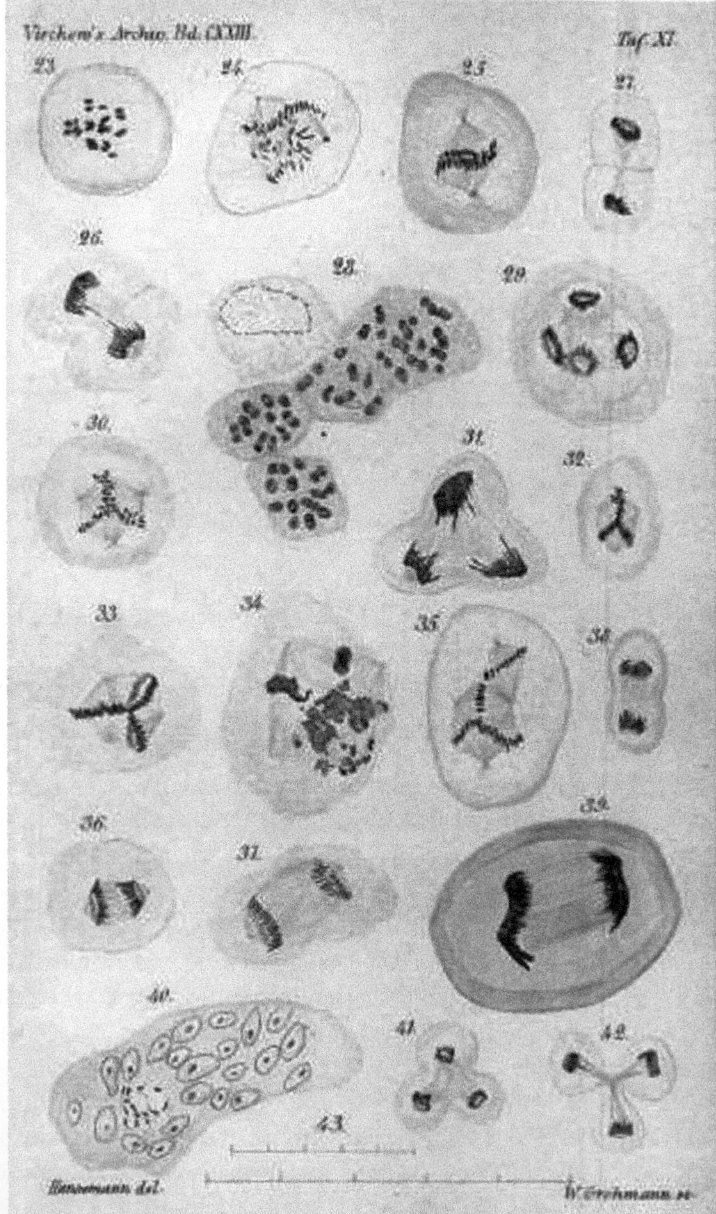

FIGURE 4.1 First observation of irregular mitoses in cancer cells as drawn by David Hansemann (1890). *Virchows Arch A Pathol Anat* 119:299–326, 1890.

a special organ of cell division. All this tied in with Boveri's interest in the chromosome theory of inheritance, referred to in the previous chapter, and, in particular, with his other but equally theoretical interest, the importance of chromosomes in cancer development.

Boveri acknowledged with gratitude the early role of Hansemann as his predecessor in focusing on the role of chromosomes in tumorigenesis, contributions that in 1900 earned Hansemann a knighthood, allowing him to add a "von" to his surname, a small change perhaps but one that has caused quite a bit of later confusion for us who have referenced his work. But apart from that, Boveri was fully cognizant of the novelty of his own thinking and the suggestions he put forward in a field to which he was alien by education as well as training. Already in 1902, Boveri had reasoned that a cancerous tumor develops from a single cell whose chromosomes have been scrambled, as the result of which it begins to divide in an uncontrolled manner. The tumor problem was thus a cell problem, not a tissue problem. However, his most famous contribution dates from 1914, one year before he passed away, when his book *Zur Frage der Entstehung maligner Tumoren* was published. In 1929, an English translation of it, undertaken by his American widow who was also a biologist, came out. Yet another translation, by Henry Harris and accompanied by his excellent annotations, was published in 2008.

We concur with Harris who in his preface called Boveri's monograph "a masterpiece of astonishing originality and foresight." Drawing on his experience with microscopic studies of nonneoplastic but sometimes chromosomally injured cells from a completely different species, Boveri exercised what must have been a very considerable ability in synthetic thinking to suggest a series of attributes of cancerous tumors that he saw as revealing of how the new growths, the neoplasms, developed from previously normal human somatic cells. Key to his thinking was the idea that malignancy is brought about by a disorder in the chromosome constitution of the cell, that (we again quote from Harris) "karyological disorder is initiated most commonly by abnormalities of mitosis and that centriolar malfunction might sometimes be involved" (Figure 4.2). Other hypotheses of Boveri's in what became known as the *somatic mutation theory of cancer* were that malignant tumors are clonal (i.e., they develop from a single, transformed mother cell), that cells tend to multiply exponentially unless restrained, that such restraint is imposed by differentiation, and that malignancy results when the said restraint is lifted by whatever mechanism (Figure 4.3).

No theory has had a more massive impact on the field of cancer cytogenetics than Boveri's. His 100-year-old predictions describe how previously healthy somatic cells escape controls of their inherent tendency to divide

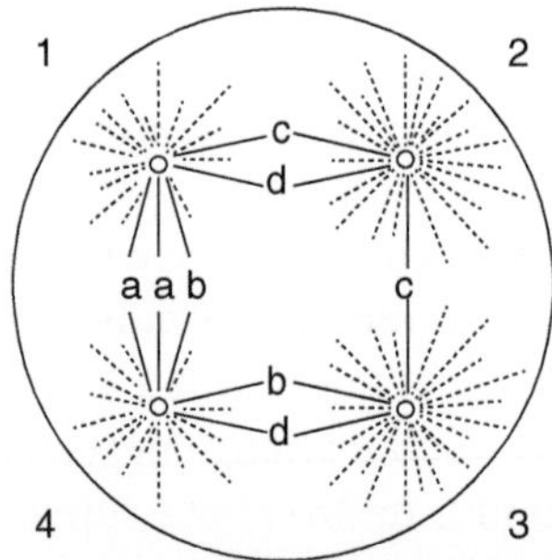

FIGURE 4.2 Boveri's illustration in *Zur Frage der Entstehung maligner Tumoren* of irregular distribution of chromosomes to daughter cells due to a tetrapolar mitosis caused by centriolar malfunction.

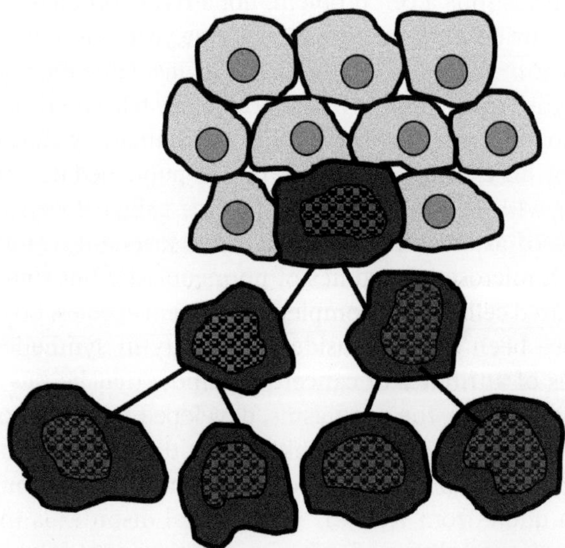

FIGURE 4.3 This highly schematic illustration has been used for 50 years when teaching medical students. It illustrates the central tenet of Boveri's somatic mutation theory of cancer: Neoplasia occurs when a suitably primed somatic cell is hit by a transforming genetic change. This defining alteration, which is often visible as a chromosomal aberration, is in later cell divisions propagated to all daughter cells, letting them grow into a karyotypically abnormal clone.

(and, in more recent versions, also their rate of death), resulting in tumor growth. Implicit in it are also several modern concepts including the primary role of "restraints" (we would today use the term "tumor suppressor genes") as well as the existence of their opposites, activated oncogenes, the importance of which will be alluded to repeatedly in later chapters.

The somatic mutation theory of cancer has stood the test of time, observation, testing, and experimentation. With as much certainty as can reasonably be asked for in biology, the gist of Boveri's suggestions is now known to be the way things are in this field. Mutation(s) acquired by somatic target cells capable of responding by excessive net division (an increased rate of cell birth minus the rate of cell death) represents the central event during tumorigenesis. This is regardless of whether the injury to the cell's genetic material is stochastic or brought about by harmful exposure to, for example, irradiation or clastogenic substances (chemicals, such as cytostatics, capable of damaging the structure of chromosomes). Boveri focused on stable damages visible at the chromosomal level, the only option available to him given the state of technology at that time. Today, one would probably have referred to them as *genomic* (the genome is the total genetic material of a given cell), but with many of the relevant changes being visible at the cytogenetic resolution level as we shall later point out repeatedly. Regardless, the central tenet of his theory remains rock solid.

At the end of this brief story about Boveri and his contributions to cancer cytogenetics, we would like to underscore two aspects: That he came to the field as a complete outsider and that his reasoning was almost exclusively theoretical (albeit based on experiments from another area of research). Such individuals often have a hard time making their voices heard within an often tightly knit community of scientists acutely aware of the perceived need to avoid interference by "amateurs." And yet, Boveri's example is far from unique when it comes to who turns out to have lasting impact in science and who does not. The difficult thing, of course, is to assess theories promoted by outsiders with the necessary curiosity and fairness, but without getting bogged down by paying attention to a succession of cranky suggestions. As far as this is concerned, little has changed during the 100 years since Theodor Boveri laid the theoretical foundations for modern-day cancer cytogenetics. And as far as "speculative" theories are concerned, then, as now, interesting hypotheses, especially those that can be tested, often turn out to be far more valuable than are trivial data even if the latter may be boringly correct.

REFERENCES AND FURTHER READING

Boveri, T. (1914). *Zur Frage der Entstehung maligner Tumoren.* Jena: Gustav Fischer.

Hansemann, D. (1890). Über asymmetrische Zelltheilung in Epitelkrebsen und deren biologische Bedeutung. *Virchows Arch. A Pathol. Anat.* 119: 299–326.

Hardy, P.A. and Zacharias, H. (2005). Reappraisal of the Hansemann–Boveri hypothesis on the origin of tumours. *Cell Biol. Int.* 29: 983–992.

Harris, H. (2008). Concerning the origin of malignant tumours by Theodor Boveri. *J. Cell Sci.* 121: 1–84.

Heim, S. (2014). Boveri at 100: Boveri, chromosomes and cancer. *J. Pathol.* 234: 138–141.

Maderspacher, F. (2008). Theodor Boveri and the natural experiment. *Curr. Biol.* 18: 279–286.

Moritz, K.B. and Sauer, H.W. (1996). Boveri's contributions to developmental biology – a challenge for today. *Int. J. Dev. Biol.* 40: 27–47.

Mukherjee, S. (2010). *The Emperor of all Maladies: A Biography of Cancer.* New York: Scribner.

CHAPTER 5

Cytogenetics from 1914 to 1960: Slow Progress Followed by Serendipitous Methodological Breakthroughs Leading to Important Discoveries

In the first half of the last century, cytogenetic studies were mostly performed on plant or insect cells. Methods were improved and many new discoveries were made. By comparison, analyses of vertebrate, let alone human, chromosomes lagged far behind.

Two particularly important technical breakthroughs in chromosome studies – the squash preparation method and colchicine pretreatment – deserve separate mention; both of them were invented by plant cytologists. Before John Belling in 1921 introduced the squash technique, researchers had relied on sections of formalin-fixed, paraffin-embedded specimens for their studies. For the examination of tissue organization during health and disease (histology and histopathology), this 100-year-old method more than sufficed – indeed, it remains indispensable even today – but for observations of single cells it had several disadvantages. First and foremost,

Abnormal Chromosomes: The Past, Present, and Future of Cancer Cytogenetics.
Sverre Heim and Felix Mitelman.

sectioning is likely to cut a chromosome into two or more parts, and even if serial sections are examined, errors are bound to occur. The squash technique eliminates these defects: The cells, including their chromosome complements, remain intact, allowing, at least in principle, the counting of chromosomes and even their individual inspection. Moreover, the pressure applied during the squashing of a suspension of cells onto a slide (usually by thumb) flattens them, leaving all chromosomes lying on a single plane of focus. Additionally, the pressure forces chromosomes to separate from one another, giving better spreads.

The second technical improvement was the addition of colchicine, an alkaloid from the bulbs of the Mediterranean plant *Colchicum*, to growing tissues prior to fixation. The effect of this step was described during the late 1930s in several studies by A.G. Avery, A.F. Blakeslee, P. Gavaudan, and A. Levan. When a cell enters mitosis under the influence of colchicine, no intranucleic spindle forms, leaving the cell arrested in mid-metaphase. Colchicine treatment thus increases the number of mitotic cells whose chromosomes are contracted enough to make them suitable (after proper staining) for microscopic examination.

However, many methodological problems remained, and for human cytogenetics in particular they long seemed unsurmountable. Mammalian chromosomes tend to crowd in the metaphase equatorial plane after dropping onto microscopic slides, appearing glued together so that there is no way to count them, let alone examine their individual morphology with any degree of reliability. Another difficulty in those early days was the lack of access to suitable materials, i.e., cells that readily entered division. In plants, this was not much of an issue since meristems always contain a good number of mitoses, and insects, ganglia, and spermatogonia likewise provided a plentiful supply of division figures. But for human studies, it appeared that the only material capable of providing a reasonable number of divisions was the testes, samples of which were not easy to obtain.

Some cytologists tried other materials and approaches. For example, T. Kemp did a considerable amount of work on human chromosomes from tissue-cultured cells in the 1920s, whereas L.F. La Cour in the 1940s attempted to study human chromosomes identified in cells from bone marrow smears. But these innovative deviations from the classical studies of sectioned testicular tissue, in which spontaneously dividing germline cells could be seen, had little impact on human and mammalian cytogenetics. Therefore, whereas adequate procedures for analyzing plant and insect chromosomes became well established, most efforts spent on mammalian cytogenetics left little or no information of lasting importance. In hindsight, it is evident that almost all the data thus generated were unreliable. T.C. Hsu thus had every reason to refer to the first half of the twentieth century as "the Dark Age" of cytogenetics.

A brief account of the many studies of human chromosomes, in particular attempts to determine their correct number, may illustrate how the field developed during that early period. From the 1890s to the 1920s, the chromosome number in humans was reported to be from 8 to over 50, although the most frequent counts were 24 for diploid cells and 12 for haploid. Hans von Winiwarter was one of the outstanding cytologists of that era. In 1912, he reported seeing 47 chromosomes in human spermatogonia and 48 in oogonia, concluding that the sex determination mechanism in our species was XX/X0. The discrepancy between von Winiwarter's count of 47/48 and the generally prevailing belief that 24 was the right number caused much consternation, some of which was dispelled by the later work of T.S. Painter who concluded that the diploid number of humans was in the forties rather than twenties. At first, Painter was uncertain whether the human diploid chromosome number was 48 or 46 but favored the latter because "in the clearest equatorial planes so far studied only 46 chromosomes have been found." However, based on studies of spontaneously dividing cells identified in testicular tissue removed from "volunteers from the Texas State Penitentiary" (one cannot but wonder what the consent procedure was like), in 1923 he revised his opinion about the correct number from 46 to 48. He also insisted that the sex determination system in humans was XX/XY instead of the XX/X0 proposed by von Winiwarter. All subsequent studies and papers from the next few decades confirmed Painter's 1923 count, and so the chromosome number of man became accepted by unanimous decision among researchers to be 48. Only in 1956 did this truth change into untruth.

In the first years after the end of the Second World War, a series of technical improvements helped lift cytogenetics onto a totally new level. Cell culturing became more reliable than before, after antibiotics and standardized, high-quality media came into common usage, colchicine-induced arrest of dividing cells increased the number of mitoses available for study, and, above all, hypotonic treatment of cells before fixation led to much improved separation of chromosomes in squash preparations (Figure 5.1). The latter "hypotonic miracle" – the use of water instead of isotonic solutions to prevent clumping of chromosomes, thus allowing observation of them individually – was no doubt the single most important factor, independently reported in 1952 by T.C. Hsu in Texas, S. Makino in Sapporo, and A. Hughes in Cambridge, England. Hsu is generally credited in the relevant literature with the discovery, probably for the following reasons: Although Hughes analyzed

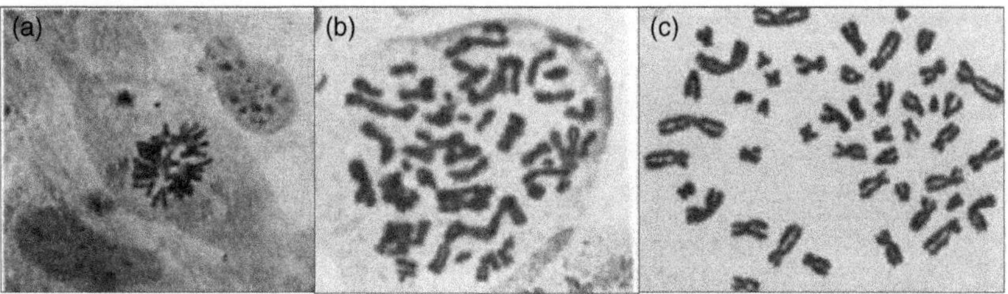

FIGURE 5.1 Human chromosome morphology showing the dramatic effect of combined treatment with colchicine and hypotonic solution. (a) Without colchicine, without hypotonic treatment (Kemp 1929). (b) With colchicine, without hypotonic treatment (Hsu 1952). (c) With colchicine and hypotonic treatment (Tjio and Levan 1956).

the effects of tonicity on chick embryonic cells *in vitro* when he found that hypotonic solutions spread the chromosomes of metaphase plates, he did not pay much attention to the observation which remained a mere by-product of his research. The article by Makino likewise did not receive much recognition, probably because he looked at tissues *in situ*, which made the chromosome dispersion effect less dramatic than when single cells are exposed *in vitro* to hypotonic solutions.

Hsu has given a vivid account of how the discovery in his laboratory was due to fortunate chance alone; it was a truly serendipitous event. Some chromosome preparations had mistakenly been rinsed in water instead of the prescribed physiological salt solution, resulting in a better spread of mitotic figures. The technician who had erred, and thus contributed so significantly to the development of cytogenetics, could not be identified and forever remains the unsung hero or heroine of the story.

As pointed out by Peter Harper, it should be added to the above account of the hypotonic miracle that similar treatments of mitotic cells apparently had been tried in the Soviet Union as early as 1934, namely by P. Zhivago and coworkers. Simultaneously and in the same country, G.K. Chrustschoff and E.A. Berlin noticed that autolysis of red blood cells tended to induce cell division in lymphocytes present in the same sample (see more below about the use of PHA for this purpose). However, because communications between the communist East and noncommunist West were at a low point during those times for both political and cultural-linguistic reasons, these discoveries remained unknown to most non-Soviet scientists. Sadly, Russian biology, and genetics in particular, which the two mentioned examples demonstrate must have been rather advanced in several areas, suffered badly from politically motivated purges in the 1930s and onward when Lysenkoism gained Stalin's favor. As a consequence, scientific progress in Russia, and hence also the rest of the world, was unnecessarily delayed.

It came like a bolt out of the blue when Joe Hin Tjio and Albert Levan in early 1956 claimed that humans did not have 48 chromosomes after all, the diploid number everybody "knew" to be correct since Painter's investigations in the 1920s and subsequently repeatedly confirmed, but 46. Tjio and Levan's study was published in *Hereditas*, the little-known journal of the Mendelian Society of Lund that at the time was primarily a forum for Swedish scientists. Thus began the modern era of human cytogenetics.

What actually happened prior to the publication of 46 as the correct chromosome number of humans has been heavily disputed, with highly divergent and even irreconcilable assessments being put forward as to the roles played by the two scientists. Since the two authors of this book met and worked together in Lund, Sweden, where the 2n = 46 research had taken place, and particularly since one of us (FM) had Albert Levan as a mentor and knew him well through many years, we would like to give a more in-depth, even personal account of that conflict than we are able to do for many of the other controversies alluded to above.

Tjio and Levan had collaborated intermittently for many years (Tjio usually stayed for weeks or months on end as a visiting scientist in Levan's laboratory) on botanical cytogenetic projects. The 2n = 46 discovery was made during such a visit between December 19, 1955 and January 26, 1956 (the same day that the paper was submitted to *Hereditas*). Toward the end of the 1950s, Tjio moved to the US and from then on claimed that he was the one who first made the correct count, at 2 a.m. on December 22, 1955 while Levan was asleep at home, and that he should have been the sole author of the paper. Levan, a shy and very modest man, always maintained that the work had been a collaborative effort. Although he never publicly criticized Tjio's view, in private conversations he expressed deep-felt sadness over what he felt was Tjio's unfair and misleading narrative. Levan's story of the events was documented (Arnason 2006) in a letter (dated Houston, Texas, May 23, 1960) written to his wife Karin while Albert was on a sabbatical in that city. We do not know why he chose to relay this story in a personal letter but probably he had, during his visit to the US, become aware of Tjio's version of what had happened 4 years previously and felt a need to make sure that at least his wife knew the truth.

In the letter, Levan told that when Tjio had arrived in Lund in December 1955, he demonstrated to him the technique he (AL) had used successfully to study chromosome preparations of human cell lines at the Memorial Sloan Kettering Institute in New York that summer (Tjio, on the other hand, had not worked on cell cultures before). He further impressed on him how essential it was to use hypotonic treatment, something that Tjio initially was reluctant to accept. Levan then wrote [. . .] "As soon as he started to follow my instructions the quality of his preparations improved and we soon sat there counting chromosomes. One evening I actually repeatedly counted 46, and the next morning I said to Hin 'Perhaps man has 46 chromosomes!' I said this more as a joke because I couldn't believe it myself." Levan continued: [. . .] "A couple of weeks passed. Then one day Hin had obtained several preparations of such a good quality that there was no doubt any longer. We counted 46 the whole day and I did a number of analyses, which we subsequently used in our publication." Now that both were convinced of the results, [. . .] "He [Tjio] suddenly asked me if he could publish the results in a Spanish journal" (Tjio was at the time on leave from an institute in Zaragoza, Spain). Levan continued in his letter: [. . .] "I refused to let Hin be the sole author of the paper, essentially because that would have been an unacceptable scientific distortion. Considering that the whole work on the human chromosomes was my plan from the beginning and that I alone had made all the preparatory work, I felt (and still feel) that I should self-evidently be the first author. I understood, however, from Hin's odd behavior (begging interrupted by outbreaks of crying) how important the issue was to him and accepted to put his name first when later that afternoon I wrote the first draft of the manuscript."

Arnason (2006) scrutinized all available historical documents related to the discovery, in particular the "logbook" kept at the Institute of Genetics in Lund, as well as records and diaries of persons who were active at the Institute around that time. His firm conclusion is that Tjio's narrative of the temporal details is incorrect. Levan seems to have made his first preliminary 2n = 46 counts around December 20–23, 1955 whereas Tjio made his first conclusive preparations in early January 1956.

The 2n = 46 discovery was made on cells cultured from lung tissue obtained from legally aborted fetuses. The photographs of the chromosome preparations, obtained by the squash technique and combined colchicine and hypotonic treatment, were of excellent quality (Figure 5.2), leading Levan to comment, many years later, that "even a child could count to 46 chromosomes." This notwithstanding, Tjio and Levan were very much aware that 48 "should" be the right number, and so they worded their different conclusion extremely carefully: "Before a renewed, detailed control has been made of the chromosome number in spermatogonial mitoses of man we do not wish to generalize our present finding into a statement that the chromosome number of man is 2n = 46, but it is hard to avoid the conclusion that this would be the most natural explanation of our observations."

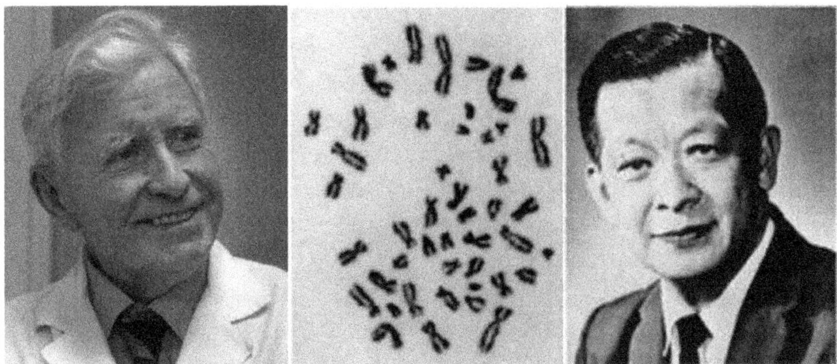

FIGURE 5.2 Albert Levan (left; 1905–1998) and Joe Hin Tjio (right; 1916–2001) with a metaphase from their publication in 1956 (*Hereditas* 42:1–6) that established the correct chromosome number of humans.

They thereby recognized the worrying possibility that the chromosome number might vary from cell type to cell type, a caveat that may seem far-fetched today but not then, considering that marked inconsistencies among tissues had been reported by several researchers over the years. It was therefore reassuring, as acknowledged in their article, that a count of 46 had repeatedly been found in an earlier study of fetal liver cells, performed by E. Hansen-Melander and her husband Y. Melander at the same institute in Lund the preceding summer. That study had been discontinued because the researchers were unable to find all the 48 chromosomes they knew must be present in the cells examined!

It did not take long before the sought-after meiotic check was performed. In November 1956, Charles Ford and John Hamerton reported 23 bivalents in the spermatocytes of all three men examined. Also other reports now corroborated the findings of Tjio and Levan proving that counting to 46 was indeed as simple as Levan later claimed, at least when no pressure was on to arrive at a different number. As a result, the correct human chromosome number was finally established. It is interesting to note, however, that the importance of this discovery was not immediately appreciated outside the small group of mammalian cytogeneticists. It was only when Jérôme Lejeune, Marthe Gautier, and Raymond Turpin in 1959 reported that three individuals with Down syndrome carried a constitutional trisomy for a small autosome (later shown to be chromosome 21) that the significance of the normal chromosome complement became recognized within the medical community. The first diagnosis of a constitutional human chromosome aberration or disease had been made.

It may be particularly pertinent, in a book devoted to cancer cytogenetics, to point out that an interest in cancer chromosomes was actually what lay behind the description of the correct normal human chromosome complement. Albert Levan, who initiated the project, began as a botanist but in the 1940s developed an interest in the chromosome aberrations of cancer cells. During that and the next decade, he and other pioneers of the field – among them J. Biesele, T. Hauschka, and S. Makino – made several important basic discoveries in cancer cytogenetics, experimental as well as conceptual. Most importantly, they provided evidence that the chromosomal variation often observed in malignant tumors made sense in Darwinian terms. Genetic variability allowed for phenotypic selection among cells, with each tumor stemline at any given time representing an adaptation to the environment consistent with the "survival of the fittest" concept.

These pioneering cancer cytogeneticists studied cell lines, usually established from cancer cells floating in malignant abdominal exudates, ascites fluid, because this was the only material to which they had access. Although the technical quality of the effusion metaphases thus obtained was good (Figure 5.3) – markedly abnormal chromosome structures could be recognized and compared among different cell lines – a major problem hampering their research was that the observed aberrations could not be identified with anything near precision. This was the main reason for Levan to start the project, supported financially by the Swedish Cancer Society, that eventually led to the description of the normal human chromosome complement. His motivation was thus not primarily an interest in what he found, the $2n = 46$ number which at that time seemed to have no obvious clinical implications, but rather a hope that knowledge about the chromosome changes typical of cancers could be translated into an understanding of what turned normal cells into malignant ones.

Although chromosomes from now on (about 1956) could be studied in technically good preparations, a major problem remained regarding which cells one had ready access to. Biopsy specimens are difficult to obtain and cell culturing was time consuming and required considerable expertise. It was therefore not practical or even feasible for many research institutions, let alone hospitals, to set up the necessary facilities just for the study of chromosomes. However, yet another serendipitous discovery was about to be made, by Peter Nowell in 1960, namely of how peripheral blood lymphocytes could be cultured simply and reproducibly.

Nowell was a pathologist at the University of Pennsylvania who studied cultured leukocytes as part of a leukemia project. He used the then standard method to separate erythrocytes from leukocytes by agglutinating the

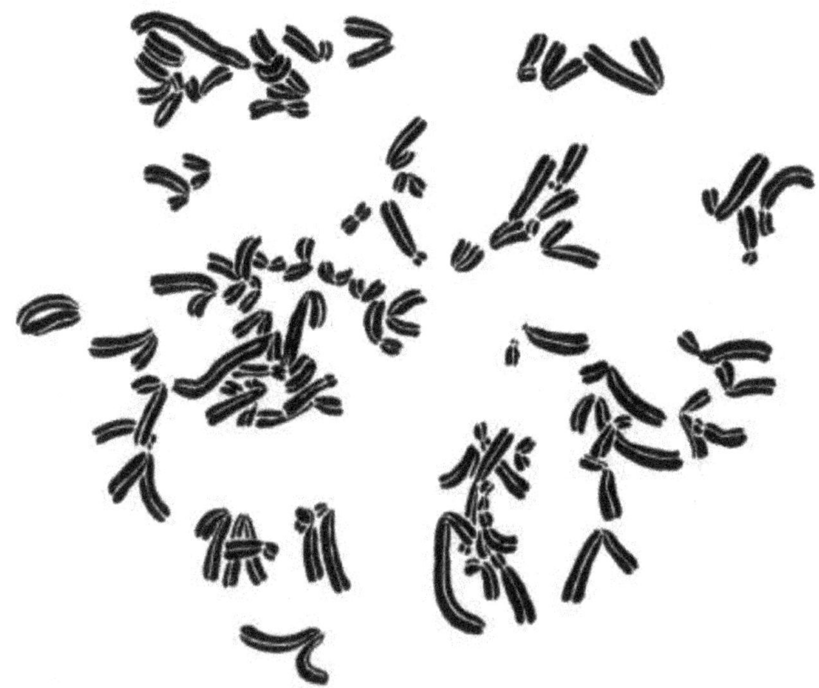

FIGURE 5.3 *Camera lucida* drawing of a tumor cell mitosis from one of the first (early 1950s) human effusions submitted to detailed chromosome analysis. Numerous abnormal chromosome shapes can be seen. Source: Courtesy of Professor Albert Levan.

red blood cells with phytohemagglutinin (PHA), a crude powder extracted from the common navy bean, *Phaseolus vulgaris*. One Friday evening, Nowell found himself running late for a family obligation and left PHA-treated specimens overnight rather than performing centrifugation after 45 minutes. When inspecting the cultures the following Monday morning, i.e. after days of incubation, he observed something highly unexpected: They contained a lot of mitotic cells. Many of the cells looked like lymphoblasts instead of terminally mature lymphocytes that do not divide. He then proceeded to test systematically all culture variables he could think of, but failed to find what had brought about the change. Since culture conditions *per se* evidently were not responsible for stimulating lymphocytes to divide, other factors must be responsible and the explanation turned out to be the PHA added. PHA not only agglutinates erythrocytes allowing separation of the two main blood cell lineages, but after 2–3 days stimulates mature lymphocytes to return to a blastic stage when they are capable of cell division.

After Nowell submitted his manuscript to *Cancer Research*, he received the reviewers' comments, and one of them wrote: "It is an interesting observation but of no conceivable significance to science." Despite this somber verdict, his turned out to be the most cited paper for more than 20 years. For some of that time, demands for PHA were so high that suppliers depleted their stock, forcing many laboratories to prepare their own bean extracts.

The discovery of the stimulatory effect of PHA on peripheral blood lymphocytes – T-cells as it later turned out – quickly led to the description by Moorhead et al. in 1960 of a simple and highly reproducible cell culturing technique used essentially unchanged to the present time. Only a small blood sample now sufficed to yield a good number of mitoses for chromosome analysis, making constitutional cytogenetic analyses easy to perform. In the mid-1960s, another important mitogen was isolated from extracts of the pokeweed plant, PWM. This substance stimulates both T- and B-lymphocytes, although at a certain concentration level mainly the latter.

When interest in cytogenetics exploded as clinically important chromosome abnormalities were detected, it soon became clear that a standardized nomenclature was needed to describe such findings. A state of affairs where every research team used their own, home-made, specialized system or language to describe what they saw was bound to lead to grave, probably even dangerous, confusion.

The pioneers in the field, the 14 scientists who had already published human karyotypes, met in Denver in 1960 to agree on how to describe chromosomes. According to the resulting Denver nomenclature, autosomes were arranged primarily according to size and numbered consecutively from 1 (largest) to 22 (smallest), whereas the sex chromosomes were designated by the time-honored symbols X and Y. Also the position of the centromere played a defining role, resulting in autosomal chromosomes being assigned to seven groups: chromosomes 1–3, 4–5, 6–12, 13–15, 16–18, 19–20, and the smallest, acrocentric chromosomes: 21–22. Three years later, at a conference in London, it was decided to refer to the seven groups of chromosomes by the letters A to G. The designations p for the short arm and q for the long arm of chromosomes were introduced at a subsequent nomenclature conference, in Chicago in 1966.

The study group at Denver did a good job. The proposed nomenclature system was robust, yet as detailed as the investigative techniques at the time allowed, and it remained virtually unaltered until the chromosome banding revolution came nearly 10 years later. After a very long adolescent period, cytogenetics had finally come of age.

REFERENCES AND FURTHER READING

Arnason, U. (2006). 50 years after – examination of some circumstances around the establishment of the correct chromosome number of man. *Hereditas* 143: 202–211.

Denver Conference (1960). A proposed standard system of nomenclature of human mitotic chromosomes (Denver, Colorado). *Ann. Hum. Genet.* 24: 319–325.

Ferguson-Smith, M.A. (2015). History and evolution of cytogenetics. *Mol. Cytogenet.* 8: e19.

Harper, P.S. (2006). *The First Years of Human Chromosomes. The Beginnings of Human Cytogenetics.* Banbury: Scion Publishing.

Harris, H. (1995). *The Cells of the Body. A History of Somatic Cell Genetics.* Long Island: Cold Spring Harbor Laboratory Press.

Hsu, T.C. (1952). Mammalian chromosomes in vitro. I. the karyotype of man. *J. Hered.* 43: 167–172.

Hsu, T.C. (1979). *Human and Mammalian Cytogenetics. An Historical Perspective.* New York: Springer.

Kemp, T. (1929). Über das Verhalten der Chromosomen in den somatischen Zellen des Menschen. *Z. mikr. Anat. Forsch.* 16: 1.

Moorhead, P.S., Nowell, P.C., Mellman, W.J. et al. (1960). Chromosome preparations of leukocytes cultured from human peripheral blood. *Exp. Cell Res.* 20: 613–616.

Nowell, P.C. (1960). Phytohemagglutinin: an initiator of mitosis in cultures of normal human leukocytes. *Cancer Res.* 20: 462–466.

Tjio, J.H. and Levan, A. (1956). The chromosome number of man. *Hereditas* 42: 1–6.

Wall, W.J. (2016). *The Search for Human Chromosomes: A History of Discovery.* New York: Springer.

CHAPTER 6

The First Cancer-Specific Chromosome Aberrations: Ph^1 and Others

The belated discovery of 46 as the normal chromosome number in humans immediately stimulated interest in cytogenetic studies of neoplastic disorders. However, despite the considerable technical progress that by the end of the 1950s had been made in mammalian cytogenetics generally, chromosome studies of cells belonging to the *parenchyma* of cancers (these are the truly neoplastic cells, whereas the other principal component of a neoplastic tumor, the *stroma* – the latter word means mattress in modern Greek – consists of all necessary supportive elements; the difference is akin to that between fighting soldiers on the one hand and tailors, cooks, chaplains, etc. on the other in an army) remained fraught with difficulties. It was still beyond the ability of most researchers to obtain a sufficient number of analyzable metaphase cells from leukemic bone marrows and, even more so, from primary solid tumors.

Besides efforts spent on the analysis of cancerous cells in animal models, most cytogenetic studies at the time therefore aimed at characterizing in detail the many cell lines that by now had been established from various malignancies. Among the most amenable materials in this regard were neoplastic cells floating in the effusions (*ascites*) that may occur in patients with intraabdominal cancers. By 1960, at least 50 human ascitic exudates had been examined cytogenetically and the results reported scientifically,

Abnormal Chromosomes: The Past, Present, and Future of Cancer Cytogenetics.
Sverre Heim and Felix Mitelman.

and the general conclusion was that at least half of the cases had a modal chromosome count deviating from 46 (Koller 1972; Atkin 1976; Sandberg 1980). Hence, they were cytogenetically abnormal.

In 1958, Ford et al. reported the first study of human leukemic cells in three patients, and in 1959 Baikie et al. described their cytogenetic findings in bone marrow cells from another 13 patients with various malignant hematological disorders. The first study to describe chromosomes in a series of solid tumors in humans was that of Makino and coworkers in Japan who in 1959 reported their findings in 19 malignant tumors from seven sites. By 1960, data from several laboratories had become available on about 120 human malignancies studied in direct preparations or in primary cultures, including 12 carcinoma types of different organs (mostly uterine, gastric, mammary, and ovarian carcinomas) and four leukemia subtypes. In total, 75% of the cases were shown to have an aneuploid karyotype, i.e., their chromosome number deviated from the normal 46. The general conclusions were as follows.

1. The stemline concept (the stemline is the most frequent chromosome constitution in a population of neoplastic cells, i.e., it takes into account both numerical and structural rearrangements) derived from studies of ascites cancers in mice and rats during the 1950s was shown to apply also to human malignancies.
2. A wide range of variation in chromosome number could typically be seen in each tumor (48–120). In general, however, it seemed that leukemias were usually diploid or near-diploid, whereas solid cancers often had modal chromosome numbers in the triploid-tetraploid range.
3. Different tumor types originating from the same organ often had different stemline chromosome patterns. There thus seemed to exist some sort of type-specificity among human cancers, although considerable tumor "individuality" could also be seen, with each new neoplasm being characterized by its own mode of chromosome number.
4. Morphologically altered chromosomes (i.e., they differ in size and shape from those of normal karyotypes) were observed in many tumors. The aberrations of cancers could evidently be both numerical and structural.

This was the situation when Peter Nowell and David Hungerford in 1960 reported, in a very brief letter to the Editor of *Science*, the finding of a minute

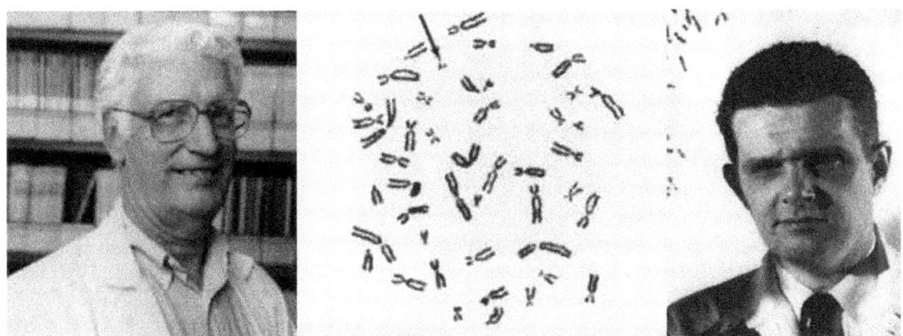

FIGURE 6.1 Peter C. Nowell (1928–2016) (left) and David A. Hungerford (1927–1993) (right). In the middle is a bone marrow metaphase from a patient with chronic myeloid leukemia. The arrow indicates the Philadelphia (Ph) chromosome (previously called Ph^1), the first consistent chromosome abnormality detected in a human malignancy. Source: Lasker Foundation and Alice Hungerford/The Scientist.

chromosome replacing one of the four smallest autosomes (Figure 6.1) in bone marrow cells from all seven examined individuals with chronic granulocytic (myeloid) leukemia (CML). Their final paragraph contained a seminal statement: "The findings suggest a causal relationship between the chromosome abnormality observed and chronic granulocytic leukemia." In another, more extensive article by Nowell and Hungerford (1960b) published almost simultaneously, the findings in the seven cases as well as in three additional patients were reported in more detail. In their total series, nine of 10 CML cases showed an apparently identical, minute, abnormal chromosome in the leukemic cells.

The observed correlation between a deleted G-group chromosome and CML was quickly confirmed by Baikie et al. (1960) in Edinburgh, and the following year, similar results were reported by several other research groups from Australia, Europe, and the US (Adams et al. 1961; Ishihara et al. 1961; Kinlough and Robson 1961; Ohno et al. 1961; Tough et al. 1961). Later, in 1967, the clonal origin of the marker-carrying cells was established by means of X chromosome inactivation studies by P.J. Fialkow's group who used glucose-6-phosphate dehydrogenase (G6PD) as a marker; in samples from female CML patients who were heterozygous at this X-linked locus, either the paternal or the maternal gene was found to be expressed, not both. Hence, the leukemia must have developed from a single, transformed precursor whose daughter cells had retained the same X inactivation pattern. The immaturity, indeed polypotency, of this precursor was inferred from the fact that also nucleus-containing erythroid cells in CML bone marrows carried the little marker, an intriguing piece of information discovered 4 years prior to the G6PD experiments (Tough et al. 1963; Truillo and Ohno 1963; Whang et al. 1963). On the other hand, repeated cytogenetic analyses of stromal cells from the often fibrotic bone marrows of patients with CML never detected the same marker or, for that matter, any other aberration. This was conclusive evidence that only the parenchyma, the truly leukemic cells, contained the acquired chromosome abnormality that presumably brought about leukemic transformation. The fibrosis was not part of the neoplastic process but a response to it.

During the course of the disease which at that time was invariably fatal, many CML patients developed additional chromosome abnormalities in their bone marrow. Serial investigations demonstrated that the karyotypic abnormalities occurring in excess of the characteristic little marker chromosome typically involved sequential acquisition of an extra marker of the same size, probably a duplication of the first one, an additional C-group chromosome, and/or the simultaneous gain of a C-group chromosome and loss of an E-group chromosome, identified as a possible isochromosome 17q. Thus, a clearly nonrandom clonal evolution pattern was established (Table 6.1).

The story told above obviously represented a resounding success. It seemed to provide decisive support for Boveri's 1914 somatic mutation theory of cancer, the supposition that acquired chromosome aberration(s) play a major role in the initiation of carcinogenesis, in this case leukemogenesis. The small, abnormal marker was soon called the Philadelphia chromosome after the city where it had been discovered, and was abbreviated Ph^1. Use of the Ph^1 symbol was originally suggested by Tough et al. (1961), the Scottish group that had been the first to confirm the marker's occurrence in CML. The superscript indicated the optimistic forecast that this was but the first consistent, even neoplasia-specific, acquired aberration found in Philadelphia; underneath lay the supposition that many more were to come. This naming tradition was soon abandoned, however; there never was a Chicago chromosome or a Lund chromosome, although the term Ph^1 was retained as a sentimental exception. Eventually, after 30 years, the superscript was in 1991 removed from the official cytogenetic nomenclature.

As already alluded to, the discovery of the Ph chromosome seemed perfect proof that neoplasia originates in one single cell which acquires a stable genetic change that not only frees it from control mechanisms whose normal function it is to prevent uncalled for cell division, but which is subsequently propagated during new mitoses to all daughter cells. It was reasonable to assume that similar discoveries would soon be made in other human cancers as well.

TABLE 6.1 Characteristic chromosome abnormalities detected in unbanded preparations in the 1960s.

Hematological disorders	
G-deletion (Ph chromosome) in CML	Nowell and Hungerford (1960a)
+Ph in Ph-positive CML	Hammouda et al. (1964), Kemp et al. (1964), de Grouchy et al. (1965), Dougan and Woodliff (1965), Kiossoglou et al. (1965b), Spiers and Baikie (1965), Erkman et al. (1966), Stich et al. (1966), Tjio et al. (1966)
+C in Ph-positive CML	Goh et al. (1964), Gruenwald et al. (1965), Kiossoglou et al. (1965b), de Grouchy et al. (1966), Lawler and Galton (1966), Stich et al. (1966), Tjio et al. (1966)
+C, −E in Ph-positive CML	Pedersen (1964), de Grouchy et al. (1966), Lawler and Galton (1966)
Iso-E in Ph-positive CML	Stich et al. (1966), Engel et al. (1968)
Ph chromosome in AML	Kiossoglou et al. (1965a), Tjio et al. (1966), Mastrangelo et al. (1967), Grossbard et al. (1968)
Ph chromosome in ALL	Propp and Lizzi (1970), Sakurai (1970)
+C in AML/MDS	Hungerford and Nowell (1962), Warkany et al. (1963), Weinstein and Weinstein (1963), Reisman et al. (1964), Sandberg et al. (1964a), Kiossoglou et al. (1965a), Rowley et al. (1966)
−C in AML/MDS	Sandberg et al. (1964b), Kiossoglou et al. (1965a), Rowley et al. (1966)
+G in AML	Sandberg et al. (1964b), Ilbery and Ahmad (1965), Kiossoglou et al. (1965a), Khan and Martin (1967), Engel et al. (1968), Goldberg et al. (1968), Sandberg et al. (1968)
C-G translocation (−C,−G,+D) in AML	Krogh Jensen (1967), Kamada et al. (1968), Ezdinli et al. (1969), Makino et al. (1969)
F-deletion in polycythemia vera	Kay et al. (1966), Lawler et al. (1970)

(continued)

TABLE 6.1 *(continued)*

Hematological disorders	
Solid tumors	
−G in meningioma	Zang and Singer (1967), Mark (1970), Singer and Zang (1970)
Double minute chromosomes in neurogenic tumors	Cox et al. (1965), Lubs and Salmon (1965), Lubs et al. (1966), Levan et al. (1968)

Abbreviations C, D, E, F, and G refer to chromosome groups 6–12, 13–15, 16–18, 19–20, and 21–22, respectively, according to the London nomenclature used in the 1960s (see previous chapter).

ALL, acute lymphoblastic leukemia; AML, acute myeloid leukemia; CML, chronic myeloid leukemia; MDS, myelodysplastic syndrome.

The boost of interest in cancer cytogenetics enjoyed after Nowell and Hungerford's discovery led to several reports in the 1960s describing a large number of cytogenetically examined leukemias, lymphomas, and various solid tumors, but the results did not meet the high hopes generally held. Chromosome abnormalities did exist in most malignancies, that much was clear, but no new, specific aberration comparable to the Philadelphia chromosome was detected. The general impression therefore spread that the aberrations detected showed no consistent pattern even among tumors of the same type. As a consequence, toward the end of the decade most scientists agreed that chromosome abnormalities were secondary epiphenomena – the consequence of neoplasia, not its cause. The Ph began to be viewed as a weird irregularity, the exception that proved the rule that chromosome change does not play any important pathogenetic role in carcinogenesis.

Nevertheless, the gloomy feeling toward the end of the 1960s that cytogenetics was unable to provide any significant key to the understanding of how cancer in humans develops was not entirely justified. Several important discoveries, listed in Table 6.1, were being made at the time, although this progress was of course completely overshadowed during the following decade when the full impact of banding cytogenetics made itself felt.

In subsets of acute myeloid leukemia (AML) and myelodysplasia (MDS; also often called preleukemia, dysmyelopoietic syndrome or smoldering leukemia at the time), some characteristic numerical aberration patterns were established involving gain of an extra C-group chromosome, loss of a C-group chromosome, trisomy for a G-group chromosome, and a C-G translocation seen together with −C, −G, and +D; the latter aberration would correspond to what the t(8;21) characteristic of AML (see next chapter) would look like in classically stained, unbanded preparations. Finally, a deletion of an F-group chromosome (later identified as 20q−) was recurrently found in bone marrow cells from patients with chronic myeloproliferative disorders, especially polycythemia vera.

Studies of solid tumors continued to lag behind those of hematological disorders, mainly for technical reasons but also because of the often remarkable complexity and heterogeneity of the abnormalities seen. Nevertheless, by the mid-1960s a partly orderly pattern had been established in even this seemingly chaotic field. Surveying chromosomal data available in the literature, mainly on ascitic forms of gastric, mammary, uterine, and ovarian carcinomas, Levan (1966) and van Steenis (1966) independently found clear evidence that certain chromosome types tended to increase and others to decrease in number in these malignancies. The nonrandomness of karyotypic changes was also demonstrated beyond doubt in several comprehensive cytogenetic studies of experimental tumors, including 7,12-dimethylbenz(a)anthracene (DMBA)-induced leukemias and sarcomas in rats (Sugiyama et al. 1967; Mitelman and Levan 1972), and sarcomas induced by the Rous sarcoma virus (RSV) in mice (Mark 1967), Chinese hamster (Kato 1968), and rats (Mitelman 1971) (Figure 6.2).

A characteristic abnormality was discovered in 1967 in benign meningiomas by Zang and Singer, namely loss of a G-group chromosome (it was later shown to be −22), proving that characteristic acquired chromosomal changes were not restricted to malignant neoplasms. Finally, double minute

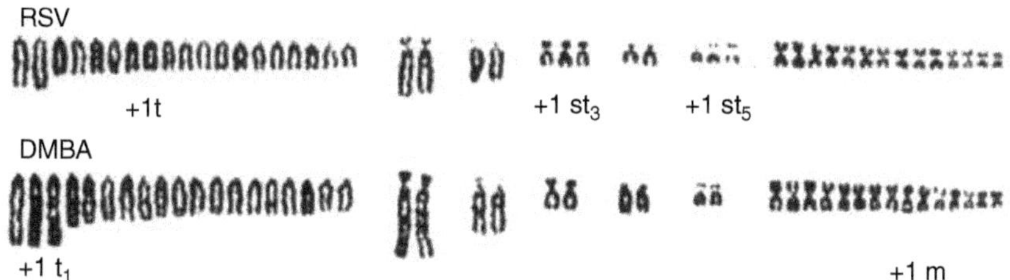

FIGURE 6.2 Karyotypes of experimental rat sarcomas induced by the Rous sarcoma virus (RSV) and 7,12-dimethylbenz(a)anthracene (DMBA) showing characteristic aberration patterns depending on the inducing agent: An additional t, st_3, and st_5 chromosome in RSV sarcomas, and an extra t_1 and m chromosome in the DMBA sarcomas. Source: Mitelman et al. (1972).

chromosomes – later shown to contain highly amplified genetic material, often of an oncogenic nature – were repeatedly found in neurogenic tumors.

The first *clinical-cytogenetic associations* were established during this time, laying a sound foundation for the extensive later use of cytogenetics as a diagnostic and prognostic tool in hematology. First and foremost, search for the presence of a Ph chromosome soon became an integral part of state-of-the-art diagnostic practices when a patient was suspected of having CML. Important not only for our understanding of the disease process, but also from a clinical point of view was that the Ph remained detectable in CML bone marrows during the whole duration of the disease, even in cases which by standard morphological-hematological investigation seemed to have entered complete remission (Sandberg et al. 1962; Carbone et al. 1963; Fitzgerald et al. 1963; Tough et al. 1963; Whang et al. 1963; Reisman et al. 1964; Frei et al. 1964).

It gradually became apparent that about 10–15% of patients with CML did not have a Ph chromosome, and several studies showed that these Ph-negative patients had a shorter survival time (Krauss et al. 1964; Tjio et al. 1966; Whang-Peng et al. 1968; Ezdinli et al. 1970). But the impact of the new genetic mode of diagnosing one of the classic leukemias had become so strong that relatively soon the inclusion criteria for making a CML diagnosis were reformulated; Ph-negative cases were moved to the chronic myelomonocytic subgroup of MDS whereas, on the other hand, Ph-positive cases of essential thrombocytosis now became thought of as actually representing CML, albeit with unusually high platelet counts.

Thus, the first truly cancer-specific acquired chromosome abnormality spearheaded a change of classification practices in the direction of paying more attention to pathogenetic differences as these gradually became apparent. The time-honored, classic disease terms were retained, however; we still talk about acute and chronic leukemias but without any longer meaning brisk onset and slow onset by these adjectives. First, how quickly the blood became white (the literal meaning of the word *leukemia*) and how fast the patient became ill had to give way to a morphological description of the leukemic cells (whether they were immature or mature in appearance corresponding to acute and chronic, respectively), then leukemia classification became increasingly dependent on genetic disease features.

The appearance of other chromosomal aberrations in addition to the Ph in CML was shown to herald imminent blast crisis (Whang-Peng et al. 1968), the disease transformation to a more acute leukemia that eventually took the patient's life. Chromosome analyses of bone marrow aspirates thus became an important aid in the follow-up of CML cases.

About 50% of acute leukemias, be they myeloid or lymphatic, were found to have an abnormal bone marrow karyotype. In contrast to what happened in CML, however, these abnormal clones disappeared in remission only to reappear during relapse (Hungerford and Nowell 1962; Sandberg et al. 1964b; Reisman et al. 1964; Kiossoglou et al. 1965a; Sandberg et al. 1968; Hart et al. 1971). This rule of thumb also held true for the relatively rare Ph-positive AML and acute lymphoblastic leukemia (ALL) cases; in them, too, the bone marrow became normal during remission, something that helped distinguish these primary acute leukemias from CMLs that made their debut in blast crisis.

The chromosome aberration pattern in diagnostic bone marrows was found to carry prognostic information, too. AML patients with only abnormal metaphases had worse prognosis than patients who had either a normal marrow karyotype or a mixture of normal and abnormal cells (Sakurai and Sandberg 1973), a finding which was very influential at the time. The conclusion has not stood the test of time after previously undetectable aberrations began to be discovered by banding analysis – which in its turn led to the detection of new cytogenetic-prognostic correlations – but the subdivision into cytogenetically defined subgroups was conceptually of considerable importance in establishing the karyotype as a possible prognosticator in AML. Later studies and refinements have confirmed this fully.

A favorable prognosis for patients with hyperdiploid ALL (life-saving treatment regimens based on the use of combinations of cytostatics had around 1960 been introduced; prior to that, all patients with acute leukemia invariably died from their disease, making it meaningless to discuss expected outcome in terms like better or worse) was shown in 1967 by F. Lampert. Investigating 10 children with ALL, he noted that the four patients whose bone marrow karyotypes had 50–59 chromosomes responded well to induction therapy and also had long survival. This very important observation was not often cited in spite of the correlation since being confirmed repeatedly in countless studies, perhaps because it was published in a non-English language journal. However, it eventually led to a new risk stratification of children with ALL according to which more aggressive treatment was being administered to high-risk patients, including those with adverse genetic leukemia features.

It may be prudent at this juncture to add a general comment on prognostic factors, including karyotypic ones: They reflect not only information on how aggressive the malignant disease is, but even more so how effective is the treatment. They are entirely empirical in nature; what yesterday was an unfavorable prognostic marker may tomorrow, after a new type of therapy has become available, become the opposite. This has already happened more than once in medical history.

A clinically important offshoot of the chromosomal analyses of cells in malignant effusions is also worthy of mention. The detection in the 1960s of highly abnormal metaphases in such materials laid the foundation for the use of cytogenetic analyses of pleural and ascitic exudates as a diagnostic aid to differentiate malignant processes from benign ones (Ishihara et al. 1961; Spriggs et al. 1962; Makino et al. 1964; Sandberg et al. 1967). Whereas the finding of a normal karyotype was inconclusive in such differential diagnostic dilemmas, the detection of highly chromosomally abnormal cells indicated with near certainty that the exudate was malignant in nature.

Another, perhaps slightly off-topic discovery also took place in the 1960s, one that only indirectly had to do with the cytogenetics of cancer: The chromosome breakage syndromes were described. Fanconi anemia was the first disease in which spontaneous chromosomal breakage was detected, both *in vitro* and *in vivo* (Schroeder et al. 1964). Other autosomal recessive conditions characterized cytogenetically by chromosomal fragility of somatic cells were Bloom syndrome (German et al. 1965) and ataxia telangiectasia (Hecht et al. 1966). In addition to their other features, patients suffering from these diseases also displayed a certain proneness to develop malignant disorders, emphasizing yet again that a pathogenetic link existed between chromosomes and cancers.

So cytogenetics was definitely far from dead toward the end of the 1960s although comments to this effect were repeatedly made by several critics both at the time and retrospectively. Chromosome analysis was not "just stamp collecting, probably fun for those involved, but of no potential theoretical or practical value." Even if no further technological advances had been made, cytogenetic analyses would probably have continued to be of use, both as a research tool and in certain clinical situations. New discoveries would most probably have followed. But all these counterfactual speculations are futile, for, as so often before, something unexpected happened. While the decade was drawing to a close, the banding revolution came and injected new life into all kinds of chromosome studies.

REFERENCES AND FURTHER READING

Adams, A., Fitzgerald, P.H., and Gunz, F.W. (1961). A new chromosome abnormality in chronic granulocytic leukaemia. *Br. Med. J.* 2 (5265): 1474–1476.

Atkin, N.B. (1976). *Cytogenetic Aspects of Malignant Transformation*. Basel: S. Karger.

Baikie, A.G., Court Brown, W.M., Jacobs, P.A., and Milne, J.S. (1959). Chromosome studies in human leukemia. *Lancet* 2: 425–428.

Baikie, A.G., Court-Brown, W.M., Buckton, K.E. et al. (1960). Possible specific chromosome abnormality in human chronic myeloid leukaemia. *Nature* 188: 1165–1166.

Carbone, P.P., Tjio, J.H., Whang, J. et al. (1963). The effect of treatment in patients with chronic myelogenous leukemia. Hematologic and cytogenetic studies. *Ann. Intern. Med.* 59: 622–628.

Cox, D., Yuncken, C., and Spriggs, A.I. (1965). Minute chromatin bodies in malignant tumours. *Lancet* I: 55–58.

Dougan, L. and Woodliff, H.J. (1965). Presence of two Ph-1 chromosomes in cells from a patient with chronic granulocytic leukaemia. *Nature* 205: 405–406.

Engel, E., McKee, L.C., and Engel-de Montmollin, M. (1968). Aberrations chromosomiques dans les maladies malignes du sang. *Union Med. Can.* 97: 901–906.

Erkman, B., Crookston, J., and Conen, P.E. (1966). Double Ph 1 chromosomes in leukaemia. *Lancet* 1 (7433): 368–369.

Ezdinli, E.Z., Sokal, J.E., Aungst, C.W. et al. (1969). Myeloid leukemia in Hodgkin's disease: chromosomal abnormalities. *Ann. Intern. Med.* 71: 1097–1104.

Ezdinli, E.Z., Sokal, J.E., Crosswhite, L., and Sandberg, A.A. (1970). Philadelphia-chromosome-positive and -negative chronic myelocytic leukemia. *Ann. Intern. Med.* 72: 175–182.

Fialkow, P.J., Gartler, S.M., and Yoshida, A. (1967). Clonal origin of chronic myelocytic leukemia in man. *Proc. Natl. Acad. Sci. U.S.A.* 58: 1468–1471.

Fitzgerald, P.H., Adams, A., and Gunz, F.W. (1963). Chronic granulocytic leukemia and the Philadelphia chromosome. *Blood* 21: 183–196.

Ford, C.E., Jacobs, P.A., and Lajtha, L.G. (1958). Human somatic chromosomes. *Nature* 181: 1565–1568.

Frei, E., Tjio, J.H., Whang, J., and Carbone, P.P. (1964). Studies of the Philadelphia chromosome in patients with chronic myelogenous leukemia. *Ann. NY Acad. Sci.* 113: 1073–1080.

German, J., Archibald, R., and Bloom, D. (1965). Chromosomal breakage in a rare and probably genetically determined syndrome of man. *Science* 148: 506–507.

Goh, K.O., Swisher, S.N., and Troup, S.B. (1964). Submetacentric chromosome in chronic myelocytic leukemia. *Arch. Intern. Med.* 114: 439–443.

Goldberg, L.S., Winkelstein, A., and Sparkes, R.S. (1968). Acquired G-group trisomy in acute monomyeloblastic leukemia. *Cancer* 21: 613–618.

Grossbard, L., Rosen, D., McGilvray, E. et al. (1968). Acute leukemia with Ph1-like chromosome in an LSD user. *JAMA* 205: 791–793.

de Grouchy, J., de Nava, C., and Bilski-Pasquier, G. (1965). Duplication d'un Ph^1 et suggestion d'une evolution clonale dans une leukémie myéloïde chronique en transformation aiguë. *Nouv. Rev. Fr. Hematol.* 5: 69–78.

de Grouchy, J., de Nava, C., Cantu, J.M. et al. (1966). Models for clonal evolutions: a study of chronic myelogenous leukemia. *Am. J. Hum. Genet.* 18: 485–503.

Gruenwald, H., Kiossoglou, K.A., Mitus, W.J., and Dameshek, W. (1965). Philadelphia chromosome in eosinophilic leukemia. *Am. J. Med.* 39: 1003–1010.

Hammouda, F., Quaglino, D., and Hayhoe, F.G. (1964). Blastic crisis in chronic granulocytic leukemia. Cytochemical, cytogenetic, and autoradiographic studies in four cases. *Br. Med. J.* 1 (5393): 1275–1281.

Hart, J.S., Trujillo, J.M., Freireich, E.J. et al. (1971). Cytogenetic studies and their clinical correlates in adults with acute leukemia. *Ann. Intern. Med.* 75: 353–360.

Hecht, F., Koler, R.D., Rigas, D.A. et al. (1966). Leukemia and lymphocytes in ataxia telangiectasia. *Lancet* 2: 1193.

Heim, S. and Mitelman, F. (1987). *Cancer Cytogenetics*. New York: Alan R. Liss.

Hungerford, D.A. and Nowell, P.C. (1962). Chromosome studies in human leukemia. III. Acute granulocytic leukemia. *J. Natl. Cancer Inst.* 29: 545–565.

Ilbery, P.L. and Ahmad, A. (1965). An extra small acrocentric chromosome in a case of acute monocytic leukemia. *Med. J. Aust.* 2: 330–332.

Ishihara, T., Moore, G.E., and Sandberg, A.A. (1961). Chromosome constitution of cells in effusions of cancer patients. *J. Natl. Cancer Inst.* 27: 893–933.

Kamada, N., Okada, K., Ito, T. et al. (1968). Chromosomes 21-22 and neutrophil alkaline phosphatase in leukaemia. *Lancet* 1: 364.

Kato, R. (1968). The chromosomes of forty-two primary Rous sarcomas of the Chinese hamster. *Hereditas* 59: 63–119.

Kay, H.E.M., Lawler, S.D., and Millard, R.E. (1966). The chromosomes in polycythemia vera. *Br. J. Haematol.* 12: 507–527.

Kemp, N.H., Stafford, J.L., and Tanner, R. (1964). Chromosome studies during early and terminal chronic myeloid leukemia. *Br. Med. J.* 1 (5389): 1010–1014.

Khan, M.H. and Martin, H. (1967). G 21 trisomy in a case of acute myeloblastic leukaemia. *Acta Haematol.* 38: 142–146.

Kinlough, M.A. and Robson, H.N. (1961). Study of chromosomes in human leukaemia by a direct method. *Br. Med. J.* 2 (5259): 1052–1055.

Kiossoglou, K.A., Mitus, W.J., and Damashek, W. (1965a). Chromosomal aberrations in acute leukemia. *Blood* 26: 610–641.

Kiossoglou, K.A., Mitus, W.J., and Dameshek, W. (1965b). Two Ph1 chromosomes in acute granulocytic leukaemia. A study of two cases. *Lancet* 2 (7414): 665–668.

Koller, P.C. (1972). *The Role of Chromosomes in Cancer Biology*. Berlin: Springer Verlag.

Krauss, S., Sokal, J.E., and Sandberg, A.A. (1964). Comparison of Philadelphia chromosome-positive and -negative patients with chronic myelocytic leukemia. *Ann. Intern. Med.* 61: 625–635.

Krogh Jensen, M.K. (1967). Chromosome studies in acute leukaemia. II. A comparison between the chromosome patterns of bone marrow cells and cells from the peripheral blood. *Acta Med. Scand.* 182: 157–165.

Lampert, F. (1967). Cellulärer DNS-Gehalt und Chromosomenzahl bei der akuten Leukämie im Kindesalter und ihre Bedeutung für Chemotherapie und Prognose. *Klin. Wschr.* 45: 763–768.

Lawler, S.D. and Galton, D.A. (1966). Chromosome changes in the terminal stages of chronic granulocytic leukaemia. *Acta Med. Scand. Suppl.* 445: 312–318.

Lawler, S.D., Millard, R.E., and Kay, H.E. (1970). Further cytogenetical investigations in polycythaemia vera. *Eur. J. Cancer* 6: 223–233.

Levan, A. (1966). Non-random representation of chromosome types in human tumor stemlines. *Hereditas* 55: 28–38.

Levan, A. (1967). Some current problems of cancer cytogenetics. *Hereditas* 57: 343–355.

Levan, A., Manolov, G., and Clifford, P. (1968). Chromosomes of a human neuroblastoma: a new case with accessory minute chromosomes. *J. Natl. Cancer Inst.* 41: 1377–1387.

Lubs, H.A. and Salmon, J.H. (1965). The chromosomal complement of human sold tumors. II. Karyotypes of glial tumors. *J. Neurosurg.* 22: 160–168.

Lubs, H.A. Jr., Salmon, J.H., and Flanigan, S. (1966). Studies of a glial tumor with multiple minute chromosome. *Cancer* 19: 591–599.

Makino, S., Ishihara, T., and Tonomura, A. (1959). Cytological studies of tumors. XXVII. The chromosomes of thirty human tumors. *Z. Krebsforsch.* 63: 184–208.

Makino, S., Sasaki, M.S., and Tinimura, A. (1964). Cytological studies of tumors. XL. Chromosome studies in fifty-two human tumors. *J. Natl. Cancer Inst.* 32: 741–777.

Makino, S., Obara, Y., Sasaki, M. et al. (1969). Cytologic studies of tumors XLVII. Acute myelogenous leukemia with C-G translocation and differential response to PHA of normal and leukemic cells of blood and marrow. *Cancer* 24: 758–763.

Mark, J. (1967). Chromosomal analysis of ninety-one primary Rous sarcomas in the mouse. *Hereditas* 57: 23–82.

Mark, J. (1970). Chromosomal patterns in human meningiomas. *Eur. J. Cancer* 6: 489–498.

Mastrangelo, R., Zuelzer, W.W., and Thompson, R.I. (1967). The significance of the Ph1 chromosome in acute myeloblastic leukemia: serial cytogenetic studies in a critical case. *Pediatrics* 40: 834–841.

Mitelman, F. (1971). The chromosomes of fifty primary Rous rat sarcomas. *Hereditas* 69: 155–186.

Mitelman, F. (1974). The Rous sarcoma virus story: cytogenetics of tumors induced by RSV. In: *Chromosomes and Cancer* (ed. J. German), 675–693. New York: Wiley.

Mitelman, F. and Levan, G. (1972). The chromosomes of primary 7,12-dimethylbenz(a)anthracene-induced rat sarcomas. *Hereditas* 71: 325–334.

Mitelman, F., Mark, J., Levan, G., and Levan, A. (1972). Tumor etiology and chromosome pattern. *Science* 176: 1340–1341.

Nowell, P.C. and Hungerford, D.A. (1960a). A minute chromosome in human chronic granulocytic leukemia. *Science* 132: 1497.

Nowell, P.C. and Hungerford, D.A. (1960b). Chromosome studies in human leukemia. II. Chronic granulocytic leukemia. *J. Natl. Cancer Inst.* 27: 1013–1035.

Ohno, S., Trujillo, J.M., Kaplan, W.D., and Kinosita, R. (1961). Nucleolus-organisers in the causation of chromosomal anomalies in man. *Lancet* 2: 123–126.

Pedersen, B. (1964). Two cases of chronic myeloid leukaemia with presumably identical 47.Chromosome cell-lines in the blood. *Acta Pathol. Microbiol. Scand.* 61: 497–502.

Propp, S. and Lizzi, F.A. (1970). Philadelphia chromosome in acute lymphocytic leukaemia. *Blood* 36: 353–360.

Reisman, L.E., Mitani, M., and Zuelzer, W.W. (1964). Chromosome studies in leukemia. I. Evidence for the origin of leukemic stem lines from aneuploid mutants. *N. Engl. J. Med.* 270: 591–597.

Rowley, J.D., Blaisdell, R.K., and Jacobson, L.O. (1966). Chromosome studies in preleukemia. I. Aneuploidy of group C chromosomes in three patients. *Blood* 27: 782–799.

Sakurai, M. (1970). Chromosome studies in hematological disorders. II. Chromosome findings in acute leukemia. *Acta Haematol. Jap.* 33: 116–126.

Sakurai, M. and Sandberg, A.A. (1973). Prognosis of acute myeloblastic leukemia: chromosomal correlation. *Blood* 41: 93–104.

Sandberg, A.A. (1980). *The Chromosomes in Human Cancer and Leukemia.* New York: Elsevier/North Holland.

Sandberg, A.A., Ishihara, T., Crosswhite, L.H., and Hauschka, T.S. (1962). Comparison of chromosome constitution in chronic myelocytic leukemia and other myeloproliferative disorders. *Blood* 20: 393–423.

Sandberg, A.A., Ishihara, T., and Crosswhite, L.H. (1964a). Group-C trisomy in myeloid metaplasia with possible leukemia. *Blood* 24: 716–725.

Sandberg, A.A., Ishihara, T., Kikuchi, Y., and Crosswhite, L.H. (1964b). Chromosomal differences among the acute leukemias. *Ann. NY Acad. Sci.* 113: 663–716.

Sandberg, A.A., Yamada, K., Kikuchi, Y., and Takagi, N. (1967). Chromosomes and causation of cancer and leukaemia. III. Karyotypes of cancerous effusions. *Cancer* 20: 1099–1116.

Sandberg, A.A., Takagi, N., Sofuni, T., and Crosswhite, L.H. (1968). Chromosomes and causation of human cancer and leukemia. V. Karyotypic aspects of acute leukemia. *Cancer* 22: 1268–1282.

Schroeder, T.M., Anschütz, F., and Knopp, A. (1964). Spontane Chromosomenaberrationen bei familiärer Panmyelopathie. *Humangenetik* 1 (2): 194–196.

Singer, H. and Zang, K.D. (1970). Cytologische und cytgenetische Untersuchungen an Hirntumoren. I. Die Chromosomenpathologie des Menschlichen Meningeoms. *Humangenetik* 9: 172–184.

Spiers, A.S. and Baikie, A.G. (1965). Chronic granulocytic leukaemia: demonstration of the Philadelphia chromosome in cultures of spleen cells. *Nature* 208: 497.

Spriggs, A.I., Boddington, M.M., and Clark, C.M. (1962). Chromosomes of human cancer cells. *Br. Med. J.* 2 (5317): 1431–1435.

van Steenis, H. (1966). Chromosomes and cancer. *Nature* 209: 819–821.

Stich, W., Back, F., Dörmer, P., and Tsirimbas, A. (1966, 1966). Doppel-Philadelphia-Chromosom und Isochromosom 17 in der terminalen phase der chronischen myeloischen Leukämie. *Klin. Wochenschr.* 44: 334–337.

Sugiyama, T., Kurita, Y., and Nishizuka, Y. (1967). Chromosome abnormality in rat leukemia induced by 7,12-dimethylbenz(a)anthracene. *Science* 158: 1058–1059.

Tjio, J.H., Carbone, P.P., Whang, J., and Frei, E. 3rd (1966). The Philadelphia chromosome and chronic myelogenous leukemia. *J. Natl. Cancer Inst.* 36: 567–584.

Tough, I.M., Court Brown, W.M., Baikie, A.G. et al. (1961). Cytogenetic studies in chronic myeloid leukaemia and acute leukaemia associated with monogolism. *Lancet* 1 (7174): 411–417.

Tough, I.M., Jacobs, P.A., Court Brown, W.M. et al. (1963). Cytogenetic studies on bone-marrow in chronic myeloid leukaemia. *Lancet* 1 (7286): 844–646.

Truillo, J.M. and Ohno, S. (1963). Chromosomal alteration of erythropoietic cells in chronic myeloid leukemia. *Acta Haematol.* 29: 311–316.

Warkany, J., Schubert, W.K., and Thompson, J.N. (1963). Chromosome analyses in mongolism (Langdon-down syndrome) associated with leukemia. *N. Engl. J. Med.* 268: 1–4.

Weinstein, A.W. and Weinstein, E.D. (1963). A chromosomal abnormality in acute myeloblastic leukemia. *N. Engl. J. Med.* 268: 253–255.

Whang, J., Frei, E., Tjio, J.H. et al. (1963). The distribution of the Philadelphia chromosome in patients with chronic myelogenous leukemia. *Blood* 22: 664–673.

Whang-Peng, J., Canellos, G.P., Carbone, P.P., and Tjio, J.H. (1968). Clinical implications of cytogenetic variants in chronic myelocytic leukemia (CML). *Blood* 32: 755–766.

Zang, K.D. and Singer, H. (1967). Chromosomal constitution of meningiomas. *Nature* 216: 84–85.

PART II

PRESENT

CHAPTER 7

The Banding Revolution: Cancer Cytogenetics in the 1970s

As with so many things in the later post-World War II phases of modern history, also for cancer cytogenetics it may be justified to claim that contemporary times, or what we call the *present* which this chapter is supposed to introduce, began in 1968. Such a statement requires a bit of explanation, so let us set forth by summarizing the state of cytogenetic affairs in the context of neoplastic disorders as the 1960s were drawing to a close.

Researchers with an interest in the subject matter – and there still were some die-hards around within the scientific community who tenaciously stuck to Boveri's belief that chromosomes held the key to an understanding of cancerous processes – knew by now that many malignancies, perhaps indeed most cases of both leukemia and malignant solid tumors, carried acquired clonal chromosome aberrations. These changes could be structural or numerical, sometimes both, but no matter what, they were viewed by cancer cytogeneticists as somehow playing an essential pathogenetic role. However, the inability to distinguish most chromosomes from one another represented a major hindrance (we recall that chromosomes were still ordered into very crude groups based on size and the position of their centromere), and so it was reasonably surmised that relevant aberrations might exist that could not yet be seen. Perhaps this was the reason why many

Abnormal Chromosomes: The Past, Present, and Future of Cancer Cytogenetics.
Sverre Heim and Felix Mitelman.

highly malignant tumors and/or leukemias seemed to be without acquired chromosome abnormalities?

Then, in 1968, an article appeared in *Experimental Cell Research* entitled "Chemical differentiation along metaphase chromosomes." Torbjörn Caspersson and Lore Zech from the Institute for Medical Cell Research and Genetics at the Karolinska Institute in Stockholm, Sweden, reported that exposure of chromosomes from the plants *Vicia faba* and *Trillium erectum*, as well as from Chinese hamster, to the fluorescing agents quinacrine dihydrochloride or quinacrine mustard resulted in an uneven uptake, imparting on the chromosomes a pattern of transverse striping or banding. Some areas or bands fluoresced more strongly, others less. Clearly, this discovery could have enormous implications for human cytogenetics if it turned out that a similar banding pattern existed distinguishing individual chromosomes in our species, and intense activities were launched toward finding such bands.

Promising results were soon obtained, almost beyond expectation. Photometric measurements of more than 5000 chromosomes were analyzed and – apart from certain minor but well-defined chromosome regions with especially strong fluorescence that could display interindividual variations – the fluorescence patterns along all chromosomes were shown to be not only distinct for each chromosome pair (1–22, X and Y), but also stable, reproducible, and the same for all examined individuals. These ground-breaking results were reported by the Stockholm group in more than 10 scientific articles published around 1970; the most widely cited paper (Caspersson et al. 1970a), usually later viewed as heralding the birth of chromosome banding, summarized the major findings. Thus began the banding revolution of human cytogenetics (Figure 7.1). From now on, individual human chromosome pairs could be distinguished, a fact that was to influence several areas of human biology and medicine.

The first chromosome banding technique was called Q-banding (after quinacrine) and relied on fluorescence in order to produce the sought-after bands. There was a drawback to this, especially for routine laboratory use, namely that the fluorescent staining quickly quenched. Quite soon, however, a number of other banding protocols were developed involving almost every agent and treatment under the sun, including several detergents, all of them with the ability to impart a banding pattern on the exposed chromosomes. The fact that most of the bands thus produced replicated the original Q-banding pattern was both a good and a bad thing. On the one hand, it indicated that the staining qualities reflected some deep, internal structural feature of chromosomes, not something stain specific, but on the other, serial use of several staining techniques on cells from the same sample did not yield new

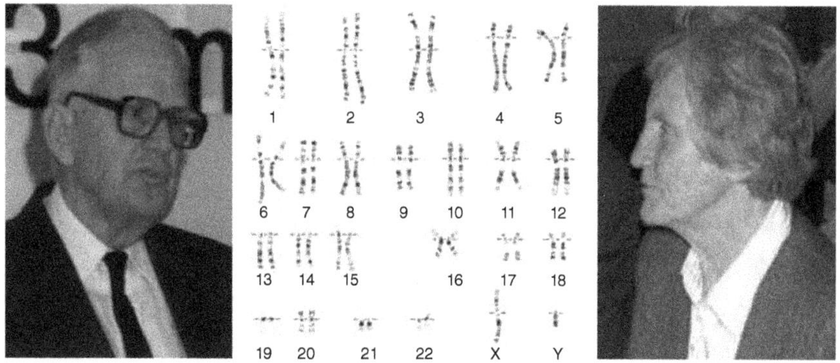

FIGURE 7.1 The pioneers of chromosome banding: Torbjörn Caspersson (1910–1997) and Lore Zech (1923–2013). A Giemsa-banded normal human male karyotype is shown in the center. The original banding technique described by Casperson and Zech was based on fluorescence staining producing Q-bands (see Figure 7.2).

information in most cases. The results did not add up; if you had performed one experiment, there was not much more to do. As usual, you win some and lose some with most changes of technique, even the most successful ones.

Some new banding procedures did have special qualities, however, inasmuch as they preferentially stained certain parts of chromosomes. What we now know as C-banding, pioneered at the turn of the decade by Pardue and Gall as well as Arrighi and Hsu, turned out to stain the densely packed heterochromatin that is mostly located in or near the centromeres of chromosomes. This approach played a role in both constitutional and cancer cytogenetics, not least in many correlation studies between the occurrence of various diseases and the amount and distribution of chromosomal heterochromatin. Seen as especially interesting at the time were the size and position of heteromorphisms connected with the large and variable heterochromatin blocks on chromosomes 1, 9, and 16.

A silver-based staining introduced in 1975 by Goodpasture and Bloom, Funaki et al., and Howell et al. had affinity for the nucleolar organizing region; it was hence called NOR- or AgNOR-banding. With its help, the short arms of acrocentric chromosomes could be identified with certainty. But more important in the rapidly growing clinical cytogenetic practice developing at the time was the introduction shortly after Q-banding of other screening but nonfluorescent techniques well suited to determining the constitutional karyotypes of patients with a syndrome-like appearance, as well as the bone marrow karyotype in leukemias. G-banding (named after its use of the Giemsa stain), developed and reported by several research groups in the early 1970s, in particular Drets and Shaw (1971), Patil et al. (1971), Schnedl (1971), and Seabright (1972), renders dark bands which are strongly fluorescent by Q-banding. R-banding gives the reverse pattern, hence its name.

It should be emphasized that none of these techniques is "better" than the others, except perhaps in very special situations; however, different labs and even countries have different traditions. For example, R-banding has been particularly popular in France, presumably because of a "founder effect"; the technique was first described by two highly influential cytogeneticists in that country, B. Dutrillaux and J. Lejeune. Examples of what chromosomes look like using the various stains are presented in Figure 7.2.

With the advent of chromosome banding, it was immediately realized that the existing nomenclature would no longer be adequate. Therefore, a group of 50 workers in the field of human cytogenetics met in Paris to agree upon a new and uniform system for the description of chromosomes and the changes they undergo. The meeting resulted in a report, entitled the Paris Conference (1971), which proved to be highly significant in the annals of human cytogenetics. It proposed a basic system for designating individual chromosomes, their regions and bands, but also for how structural rearrangements and variants should be described according to which bands they affected. Clarifications and improvements were made over the years, naturally, but the Paris guidelines to this day remain the basis for how we describe chromosomes and their abnormalities. Since the early 1980s, an International Standing Committee on Human Cytogenetic Nomenclature, elected every 5 years by the cytogenetics community, produces an updated version of the nomenclature guidelines entitled An International System for Human Cytogenetic (lately Cytogenomic) Nomenclature (ISCN).

Q-, G-, and R-banding normally give about 300 bands per haploid chromosome complement. Each of the 23 human homologue pairs displays a unique pattern of darker or lighter transverse stripes that identifies them, but also makes it possible to determine which rearrangement between them has occurred even in fairly complex aberrations.

Since the invention of chromosome banding opened up a whole new world in cancer cytogenetics, to be reviewed more extensively in the chapters to come, now may be a reasonable opportunity to summarize some of the relevant terminology used when describing changes commonly observed in neoplastic cells. For a much more thorough overview, please consult the guidelines of the latest ISCN (2020) publication. We mention here only a few particularly important terms and principles.

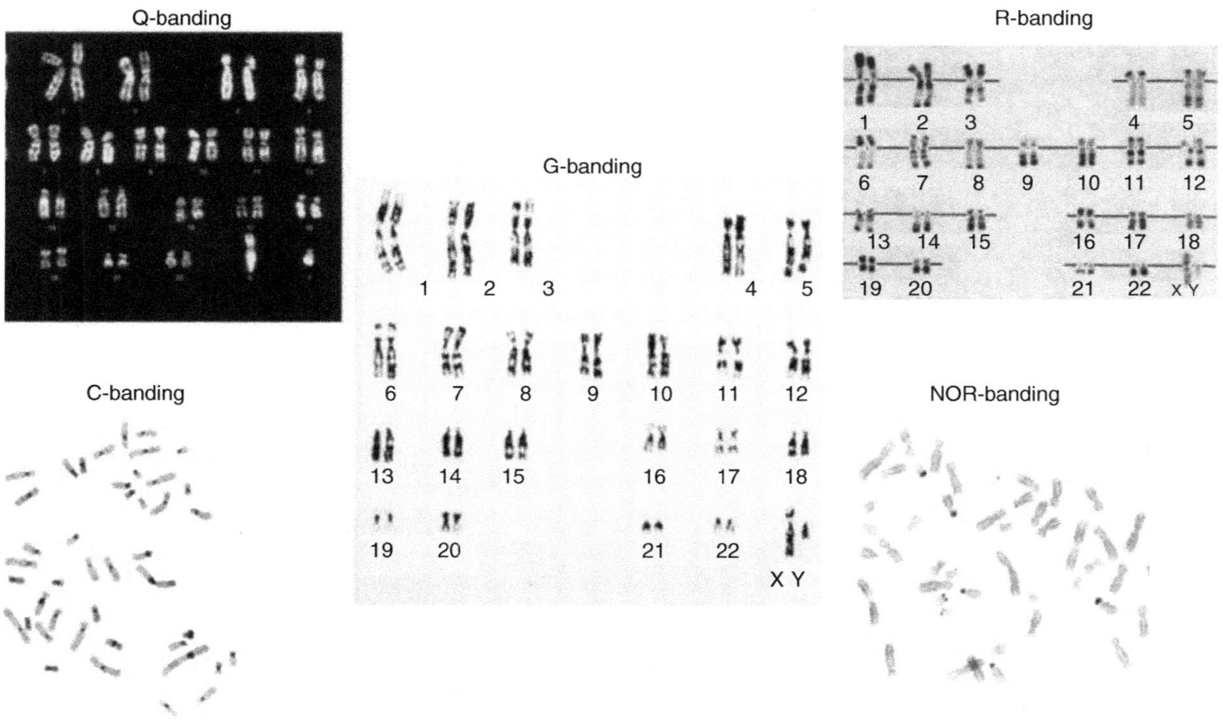

FIGURE 7.2 Examples of the various banding techniques developed during the 1970s. For details, see text.

The human chromosomes are named 1–22 in decreasing order of size, although it has later turned out that 22 is actually somewhat bigger than 21, and have one short (p) and one long (q) arm. Each arm has one or more regions (delimited by landmarks, particularly distinctive features that appear in banding preparations made by one of the standard techniques mentioned above), and within each region one sees one or more darker- or lighter-staining bands. The regions and bands are described by numbers from the centromere outwards toward the telomere. All numbering is consecutive. Thus, 9q34 means chromosome 9, the long arm, region 3 within that arm, and band 4 within that third region. Likewise, the short arm tip of chromosome 1 is called 1p36: First listed is the chromosome number, then the arm, then the region (3), and finally the band (6) within that region. For this reason, the band just mentioned is pronounced "one-p-three-six," not 36 at the end, although we are fighting a constant and losing battle in this regard with young colleagues who seem to be unaware of why things are named the way they are.

The main structural chromosome rearrangements are deletions (abbreviated *del*; a part of a chromosome arm is lost), translocations (*t*; segments are swopped between chromosomes), insertions (*ins*; a segment from one chromosome is wedged into a given site in another), and inversions (*inv*; rotation 180° of a chromosome segment). For example, del(5)(q13q33) means loss of the part of 5q between 5q13 and 5q33. A more informal and cruder way of referring to the same deletion would be to call it 5q– or del(5q) where the minus sign signifies loss, in this case from the long arm of chromosome 5 (but then no information on breakpoints is provided). t(9;22)(q34;q11) means an exchange of material between chromosomes 9 and 22 with breakpoints in 9q34 and 22q11. Finally, inv(16)(p13q22) means a 180° rotation of the part of chromosome 16 between the breakpoints 16p13 and 16q22. In rearrangements between chromosomes, a semicolon (;) is used between them, whereas no sign is used between the breakpoints of intrachromosomal rearrangements. The examples here listed are not chosen arbitrarily and will be mentioned further in later chapters.

Is it possible to obtain even more detailed banding of chromosomes than the roughly 300 bands obtained using standard techniques? If yes, would this correspond to important new information, a much more accurate picture of the aberrations thus examined or even detected at that higher resolution level? In short, would it be any good? It seems to us that the answer is both yes and no.

Many attempts were made during the 1970s and 1980s to obtain better metaphase preparations that would yield finer and more detailed banding.

In principle, two approaches were followed: Either one attempted to prevent normal shortening of the chromosomes during metaphase by adding hockey puck-shaped intercalating agents to the medium for some period of time, ensuring that the chromosomes remained long and with room for many bands, or one tried to harvest earlier during cell division (in prometaphase), when the chromosomes of most dividing cells are not yet highly condensed. Both principal methods were successful to some degree, both gave longer chromosomes with more bands, up to 3000 per haploid complement according to one of the pioneers of such fine-structure or high-resolution banding, Jorge Yunis. The additional bands that could now be seen arose by splitting up of those already known from analyses of "normal" preparations, something that became reflected in the new naming practice that became established: Each subband was given an Arabic numeral after a full stop (.) within the existing "main band." For example, 9q34.1 would mean chromosome 9, the long arm, region 3 within that arm, band 4 within that region, and subband 1 within that band.

When the time came to analyze these new fine structure preparations, however, difficulties arose, especially when the task at hand was to screen for abnormalities, i.e., when nothing was known beforehand about which specific aberration to look for. The longer and more band-rich the chromosomes became, the more tiny differences could be detected between the two homologues, but largely because of normal variation in the degree of contraction at any given point in time, not because of actual pathology. Small "deletions" were therefore sometimes "detected" where none was present. Accordingly, some exaggerated claims were made by eager scientists, claims that were later proven to be faulty.

In hindsight, one can now see that the techniques were overstretched even more than was the case with the chromosomes themselves, asked to provide more information than they could possibly deliver. Later, with the advent of fluorescence *in situ* hybridization (FISH) technologies, new and better methods were found to, amongst other things, map breakpoints to specific chromosomal positions and genes to their precise positions or loci, and so the emphasis on getting ever longer and more band-rich chromosomes for karyotyping faded into indifference.

Whether many bands or few, what chromosomal properties do they reflect? Knowledge about this is still surprisingly rudimentary, just as the relationship between structure and function at the chromosomal level remains unclear at best; we shall return to these themes on several occasions toward the end of this book. Darkly staining G-bands or their equivalent, strongly fluorescing Q-bands, tend to contain a lot of the bases adenine and thymine (AT rich) and be relatively gene poor, whereas light-staining G-bands have more guanine and cytosine (i.e., they are GC rich) and more transcriptionally active. It is therefore perhaps not surprising (to be discussed later) that cancer-associated breakpoints more often are found in light-staining (by G-banding), and hence gene-rich, genomic regions and bands.

Be all this as it may, the proof of the pudding is always in the eating, and in this regard the ushering in of chromosome banding around 1970 certainly delivered: The technical revolution led to many discoveries that also affected other areas of genetics and even medicine. Syndromology was particularly influenced; many chromosomal causes behind severe clinical conditions combining mental retardation with distinctive bodily features could now be described using banding techniques that revealed much more subtle abnormalities than the gains or losses of whole chromosomes detected during the prebanding era. In genetic research, the identification of genes and, in particular, the position of their loci within the genome took major steps forward because of the new and improved cytogenetic methods. The contributions of chromosome banding to the realization of the Human Genome Project, the sequencing of the entire genome of our species (completed in 2003) and the largest collaborative biological project so far, can hardly be overstated. Also in seemingly distant areas of research, an impact was felt, for example in the study of phylogenetic relationships between species. By means of banding studies, ancestral karyotypes could be predicted, adding significantly to the explanatory power of such studies when it came to understanding how different types of life on earth are developmentally intertwined.

But our main topic is the field of cancer cytogenetics where a whole row of important discoveries were made during the "golden age of cancer cytogenetics" that followed the introduction of Q-banding and other screening techniques from 1970 onward. More and more malignant disorders, especially leukemias, were investigated and a number of characteristic, sometimes pathognomonic, changes were soon discovered (Table 7.1). In total, more than 1200 neoplasms with clonal abnormalities were scientifically reported during this first decade of banding cytogenetics, and more than 60 recurrent chromosomal aberrations were identified.

In 1972, the nature of three of the nonrandom aberrations that had already been partially described based on examination of solid-stained preparations during the preceding decade was clarified. The additional C-group chromosome frequently detected in acute myeloid leukemia (AML) was identified as a trisomy 8 (Figure 7.3), the lost G-group chromosome of meningiomas corresponded to monosomy 22, and the deleted F-group chromosome

TABLE 7.1 Characteristic neoplasia-associated cytogenetic aberrations in human neoplasms detected by banding analyses 1970–1979.

Year	Disease	Aberration	References
1970	Chronic myeloid leukemia	del(22q)	Caspersson et al. (1970b)
1972	Acute myeloid leukemia	+8	de la Chapelle et al. (1972)
	Burkitt lymphoma	14q+	Manolov and Manolova (1972)
	Meningioma	−22	Mark et al. (1972), Zankl and Zang (1972)
	Polycythemia vera	del(20q)	Reeves et al. (1972)
1973	Acute myeloid leukemia	t(8;21)(q22;q22)	Rowley (1973a)
	Acute myeloid leukemia	i(17)(q10)	Mitelman et al. (1973)
	Acute myeloid leukemia	−7/del(7q)	Petit et al. (1973), Rowley (1973c)
	Chronic myeloid leukemia	t(9;22)(q34;q11)	Rowley (1973b)
	Acute myeloid leukemia/ myeloproliferative disorders	+9	Davidson and Knight (1973), Rowley (1973d), Rutten et al. (1973)
1974	Acute myeloid leukemia	+21	Mitelman and Brandt (1974)
	Refractory anemia/ myelodysplasia	del(5q)	van den Berghe et al. (1974)
1975	Myeloproliferative disease	t(11;20)(p15;q11)	Berger (1975)
1976	Acute myeloid leukemia	t(6;9)(p23;q34)	Rowley and Potter (1976)
	Burkitt lymphoma	t(8;14)(q24;q32)	Zech et al. (1976)
1977	Acute lymphoblastic leukemia	t(4;11)(q21;q23)	Oshimura et al. (1977)
	Acute promyelocytic leukemia	t(15;17)(q22;q21)	Rowley et al. (1977)
	Neuroblastoma	del(1p)	Brodeur et al. (1977)

(continued)

TABLE 7.1 (*continued*)

Year	Disease	Aberration	References
1978	Acute monocytic leukemia	t(8;16)(p11;p13)	Mitelman et al. (1978)
	Acute myeloid leukemia	ins(3;3)(q21;q21q26)	Golomb et al. (1978)
1979	Acute lymphoblastic leukemia	t(8;14)(q24;q32)	Berger et al. (1979a), Mitelman et al. (1979)
	Burkitt lymphoma	t(2;8)(p12;q24)	Miyoshi et al. (1979), van den Berghe et al. (1979)
	Burkitt lymphoma	t(8;22)(q24;q11)	Berger et al. (1979b)
	Chronic lymphocytic leukemia	+12	Autio et al. (1979)
	Follicular lymphoma	t(14;18)(q32;q21)	Fukuhara et al. (1979)

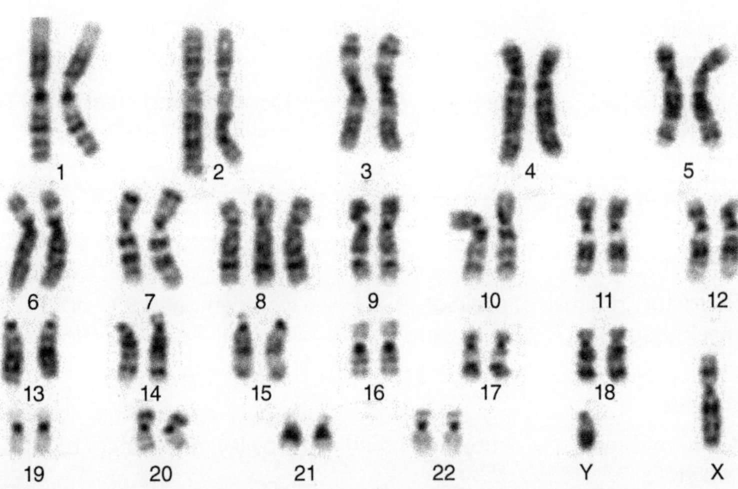

FIGURE 7.3 The first (1972) numerical abnormality in neoplasia identified by chromosome banding: Trisomy 8 in a case of acute myeloid leukemia.

sometimes seen in polycythemia vera (PV) was identified as a 20q−/del(20q). Furthermore, a previously unrecognized recurrent abnormality, a 14q+ marker chromosome occurring in Burkitt lymphoma (BL) cells, was also detected that same year.

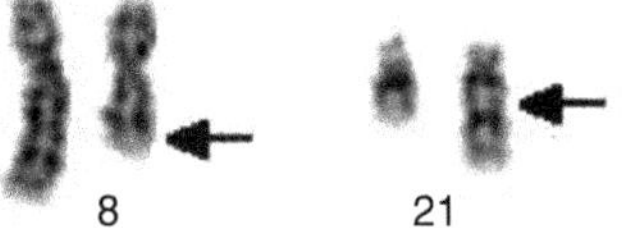

FIGURE 7.4 The first (1973) balanced rearrangement identified by chromosome banding: Translocation between chromosomes 8 and 21 in acute myeloid leukemia, t(8;21)(q22;q22). Arrows indicate the breakpoints in bands 8q22 and 21q22.

The first recurrent balanced cytogenetic rearrangements, all of them translocations, were described shortly afterwards. A reciprocal translocation between chromosomes 8 and 21, i.e., t(8;21)(q22;q22) corresponding to the C-G translocation reported previously, was found by Janet Rowley in 1973 in the bone marrow cells of two patients with AML (Figure 7.4). Around the same time, Rowley showed that the Ph chromosome of chronic myeloid leukemia (CML) originated through a t(9;22)(q34;q11), not a deletion of chromosome 22 as had been thought (Figure 7.5). Among other important

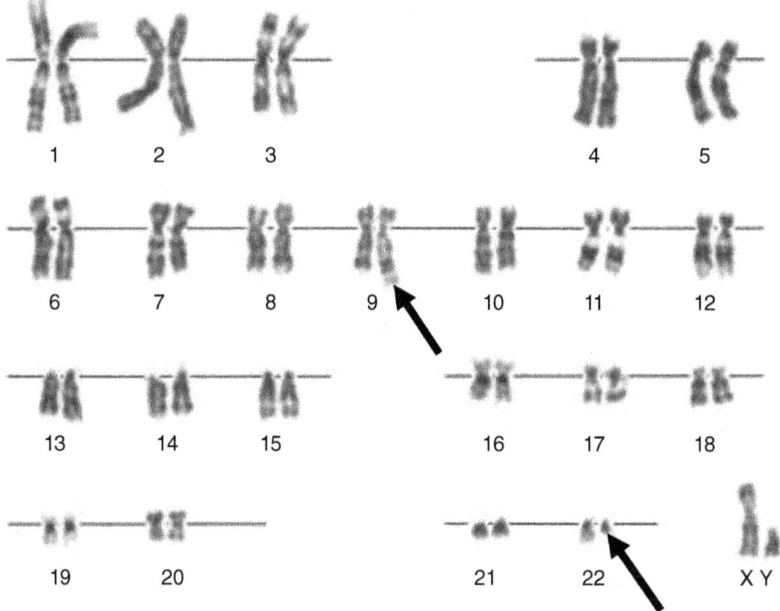

FIGURE 7.5 The balanced translocation t(9;22)(q34;q11) gives rise to the Ph chromosome that is characteristic of chronic myeloid leukemia. Until 1973, the Ph was believed to be the result of a simple deletion of the long arm of chromosome 22. The arrows point at the rearranged chromosomes.

specific translocations (Table 7.1) detected in banding preparations were t(8;14)(q24;q32), t(2;8)(p12;q34), and t(8;22)(q24;q11) in BL (the former and most common of these three gives rise to the 14q+ mentioned above), t(15;17)(q22;q21) in acute promyelocytic leukemia (APL), t(4;11)(q21;q23) in acute lymphoblastic leukemia (ALL), and t(14;18)(q32;q21) in follicular lymphoma (FL). Two characteristic translocations, T(6;15) and T(12;15) (for unknown reasons, the abbreviation is with a capital T in mice), were the first specific chromosome rearrangements found in experimental cancers, namely mouse plasmacytomas (Ohno et al. 1979). As later demonstrated (see next chapter), they turned out to be perfect pathogenetic equivalents of the above-mentioned characteristic translocations in human BL, the corresponding disease in our species. Thus began a gradual deepening of our understanding of the mechanisms that lie behind some of the most distinctive hematopoietic malignancies. More of a similar nature was to come during the next decade.

REFERENCES AND FURTHER READING

Autio, K., Turunen, O., Penttila, O. et al. (1979). Human chronic lymphocytic leukemia: karyotypes in different lymphocyte populations. *Cancer Genet. Cytogenet.* 1: 147–155.

Berger, R. (1975). Translocation t(11;20) et polyglobulie primitive. *Nouv. Press. Med.* 4: 1972.

Berger, R., Bernheim, A., Brouet, J.C. et al. (1979a). t(8;14) translocation in a Burkitt's type of lymphoblastic leukaemia (L3). *Br. J. Haematol.* 43: 87–90.

Berger, R., Bernheim, A., Weh, H.-J. et al. (1979b). A new translocation in Burkitt's tumor cells. *Hum. Genet.* 53: 111–112.

van den Berghe, H., Cassiman, J.J., David, G. et al. (1974). Distinct haematological disorder with deletion of long arm of no. 5 chromosome. *Nature* 251: 437–438.

van den Berghe, H., Parloir, C., Gosseye, S. et al. (1979). Variant translocation in Burkitt lymphoma. *Cancer Genet. Cytogenet.* 1: 9–14.

Brodeur, G.M., Sekhon, G.S., and Goldstein, M.N. (1977). Chromosomal aberrations in human neuroblastomas. *Cancer* 40: 2256–2263.

Caspersson, T., Zech, L., Johansson, C., and Modest, E.J. (1970a). Identification of human chromosomes by DNA-binding fluorescent agents. *Chromosoma* 30: 215–227.

Caspersson, T., Gahrton, G., Lindsten, J., and Zech, L. (1970b). Identification of the Philadelphia chromosome as a number 22 by quinacrine mustard fluorescence analysis. *Exp. Cell Res.* 63: 238–240.

de la Chapelle, A., Schröder, J., and Vuopio, P. (1972). 8-trisomy in the bone marrow. Report of two cases. *Clin. Genet.* 3: 470–476.

Davidson, W.M. and Knight, L.A. (1973). Acquired trisomy 9. *Lancet* 1: 1510.

Drets, M.E. and Shaw, M.W. (1971). Specific banding patterns of human chromosomes. *Proc. Natl Acad. Sci. USA* 68: 2073–2077.

Fukuhara, S., Rowley, J.D., Variakojis, D., and Golomb, H.M. (1979). Chromosome abnormalities in poorly differentiated lymphocytic lymphoma. *Cancer Res.* 39: 3119–3128.

Golomb, H.M., Vardiman, J.W., Rowley, J.D. et al. (1978). Correlation of clinical findings with quinacrine-banded chromosomes in 90 adults with acute nonlymphocytic leukemia. An eight-year study (1970–1977). *N. Engl. J. Med.* 299: 613–619.

Heim, S. and Mitelman, F. (2015). *Cancer Cytogenetics. Chromosomal and Molecular Genetic Aberrations of Tumor Cells.* Oxford: Wiley-Blackwell.

Hsu, T.C. (1979). *Human and Mammalian Cytogenetics. An Historical Perspective.* New York: Springer.

ISCN (2020). *An International System for Human Cytogenomic Nomenclature (2020)* (eds. J. McGowan-Jordan, R.J. Hastings and S. Moore). Basel: Karger.

Manolov, G. and Manolova, Y. (1972). Marker band in one chromosome 14 from Burkitt lymphomas. *Nature* 237: 33–34.

Mark, J., Levan, G., and Mitelman, F. (1972). Identification by fluorescence of the G chromosome lost in human meningomas. *Hereditas* 71: 163–168.

Mitelman, F. and Brandt, L. (1974). Chromosome banding pattern in acute myeloid leukemia. *Scand. J. Haematol.* 13: 321–330.

Mitelman, F., Brandt, L., and Levan, G. (1973). Identification of isochromosome 17 in acute myeloid leukaemia. *Lancet* 2: 972.

Mitelman, F., Brandt, L., and Nilsson, P.G. (1978). Relation among occupational exposure to potential mutagenic/carcinogenic agents, clinical findings, and bone marrow chromosomes in acute nonlymphocytic leukemia. *Blood* 52: 1229–1237.

Mitelman, F., Andersson-Anvret, M., Brandt, L. et al. (1979). Reciprocal 8;14 translocation in EBV-negative B-cell acute lymphocytic leukemia with Burkitt-type cells. *Int. J. Cancer* 24: 27–33.

Miyoshi, I., Hiraki, S., Kimura, I. et al. (1979). 2/8 translocation in a Japanese Burkitt's lymphoma. *Experientia* 35: 742–743.

Ohno, S., Babonits, M., Wiener, F. et al. (1979). Nonrandom chromosome changes involving the Ig gene-carrying chromosomes 12 and 6 in pristane-induced mouse plasmacytomas. *Cell* 18: 1001–1007.

Oshimura, M., Freeman, A.I., and Sandberg, A.A. (1977). Chromosomes and causation of human cancer and leukemia. XXVI. Banding studies in acute lymphoblastic leukemia (ALL). *Cancer* 40: 1161–1172.

Patil, S.R., Merrick, S., and Lubs, H.A. (1971). Identification of each human chromosome with a modified Giemsa stain. *Science* 173: 821–822.

Petit, P., Alexander, M., and Fondu, P. (1973). Monosomy 7 in erythroleukaemia. *Lancet* 2: 1326–1327.

Reeves, B.R., Lobb, D.S., and Lawler, S.D. (1972). Identity of the abnormal F-group chromosome associated with polycythaemia vera. *Humangenetik* 14: 159–161.

Rowley, J.D. (1973a). Identification of a translocation with quinacrine fluorescence in a patient with acute leukemia. *Ann. Genet.* 16: 109–112.

Rowley, J.D. (1973b). A new consistent chromosomal abnormality in chronic myelogenous leukaemia identified by quinacrine fluorescence and Giemsa staining. *Nature* 243: 290–293.

Rowley, J.D. (1973c). Deletions of chromosome 7 in haematological disorders. *Lancet* 2: 1385–1386.

Rowley, J.D. (1973d). Acquired trisomy 9. *Lancet* 2: 390.

Rowley, J.D. and Potter, D. (1976). Chromosomal banding patterns in acute nonlymphocytic leukemia. *Blood* 47: 705–721.

Rowley, J.D., Golomb, H.M., and Dougherty, C. (1977). 15/17 translocation, a consistent chromosomal change in acute promyelocytic leukaemia. *Lancet* 1: 549–550.

Rutten, F.J., Hustinx, T.W., Scheres, M.J., and Wagener, D.J. (1973). Acquired trisomy 9. *Lancet* 2: 455.

Sandberg, A.A. (1990). *The Chromosomes in Human Cancer and Leukemia*, 2e. St Louis: Elsevier.

Schnedl, W. (1971). Analysis of the human karyotype using a reassociation technique. *Chromosoma* 34: 448–454.

Seabright, M. (1972). Human chromosome banding. *Lancet* 1 (7757): 967.

Swansbury, J. (ed.) (2003). *Cancer Cytogenetics. Methods and Protocols.* Totowa: Humana Press.

Zankl, K. and Zang, K.D. (1972). Cytological and cytogenetical studies on brain tumors. 4. Identification of the missing G chromosome in human meningiomas as no. 22 by fluorescence technique. *Humangenetik* 14: 167–169.

Zech, L., Haglund, U., Nilsson, K., and Klein, G. (1976). Characteristic chromosomal abnormalities in biopsies and lymphoid-cell lines from patients with Burkitt and non-Burkitt lymphomas. *Int. J. Cancer* 17: 47–56.

CHAPTER 8

Chasing Correlations: Chromosomes and Oncogenes in Leukemias and Lymphomas

As already pointed out in the preceding chapter, the invention of banding methods ushered in a veritable cancer cytogenetics gold rush. The hematopoietic malignancies – lymphomas, leukemias, and various leukemia-like bone marrow neoplasias – were the most readily accessible when it came to sampling and therefore first in line to be investigated using the new techniques. This resulted in a huge amount of new and interesting information. For a complete review of that body of knowledge, see, for example, Heim and Mitelman (2015). Here, our intention is merely to highlight some of the general cytogenetic conclusions that emerged during the 1970s and 1980s, followed by examples of how the consecutive analysis of leukemic and lymphomatous cells by first cytogenetic, then molecular genetic means opened up a new understanding of how acquired chromosome abnormalities might act pathogenetically. The emphasis will be on the principles involved, which invariably means that most of the massive data accumulated (`https://mitelmandatabase.isb-cgc.org`) will go unmentioned.

In *acute myeloid leukemia* (AML), about two-thirds of all patients were found to have clonal chromosome aberration(s) in their leukemic bone

Abnormal Chromosomes: The Past, Present, and Future of Cancer Cytogenetics.
Sverre Heim and Felix Mitelman.

marrow cells at diagnosis, irrespective of which banding techniques were used. Moreover, a clearly nonrandom aberration pattern was detected. Some patients had only a single balanced structural rearrangement – t(8;21)(q22;q22), t(15;17)(q22;q11), and inv(16)(p13q22) historically have pride of place – whereas others had acquired unbalanced aberrations involving losses (often −5/5q− or −7/7q−) or gains of chromosome material (+8 was seen to be particularly frequent), either alone or in combinations of increasing complexity.

Some of the more distinct aberration patterns corresponded to particular leukemic features either hematologically or with regard to bone marrow morphology. This correspondence sometimes amounted to a strict one-to-one relationship. For example, patients with acute promyelocytic leukemia (APL) or AML M3 in the French-American-British (FAB) nomenclature (Bennett et al. 1976; the classification reflected an international attempt in the mid-1970s to subdivide AML- and ALL cases based on morphological criteria, and the leukemic granulocyte precursors of M3 patients are the most mature-looking cells you can have and still call the disease "acute") almost always had t(15;17), a translocation not found in other leukemia subsets. Some characteristic aberrations were always detected in one particular type of disease but could also be seen in other settings, albeit more rarely. The Philadelphia chromosome arising through t(9;22) in chronic myeloid leukemia marrows is historically the most prominent example. Though obligatory in that setting, the same translocation was also occasionally encountered in cases of Ph-positive myeloid and, especially, lymphocytic or lymphoblastic acute leukemia. Although the genotypic-phenotypic specificity is considerable in the case of Ph-positive leukemias, too, the diagnostic information value of detecting a t(9;22) in bone marrow cells must therefore always be viewed against the background of other relevant parameters. Even more common is that only fairly coarse quantitative differences are seen with regard to how often certain aberrations occur in particular leukemia types or subtypes; nonrandomness *is* observed, but the distribution is not skewed enough to be seen as reflecting true specificity, let alone the pathognomonic correlation between t(15;17) and APL.

Why is it that we do not always face a one-to-one relationship between pathogenetic rearrangements and phenotypically distinct malignancies, be it amongst leukemias or solid tumors (see next chapter)? The truth is that we still mostly do not know how a given genetic transforming event in a susceptible somatic cell determines the ensuing neoplasm's phenotypic features, but this ignorance should not prevent us from providing a principal answer to the question above valid in at least most cases. If we for the time being restrict our considerations to the leukemia situation, standard thinking holds that the leukemogenic abnormality is acquired by some kind of stem cell, although not necessarily one that is totipotent. If the same genetic change – again t(9;22) is a perfect example – is leukemogenic for both unipotent lymphocytic stem cells and two different types of myeloid stem cells, one of which is rigged to block differentiation early while the other is not, then the leukemic phenotype would reflect the differentiation propensity of the cell in which the translocation occurred. Perhaps t(9;22) itself contributes to arresting differentiation, perhaps not; regardless, the eventual leukemic phenotype would reflect both the type of leukemic event that occurred and the inherent abilities of the stem cell targeted. Both seed and soil are important.

The cytogenetic relationships that unfolded between the AMLs and the preleukemias often preceding them (other terms for the latter were refractory anemia, smoldering leukemia, dysmyelopoiesis and – the term mostly preferred today – myelodysplastic syndrome [MDS] or myelodysplasia) revealed close pathogenetic similarities. The aberration profiles of the two conditions displayed considerable overlap, albeit with some significant differences. Changes like +8 and 7q−/−7 were frequently seen in both MDS and AML. A large deletion of the long arm of chromosome 5 was more common in MDS than AML, especially as the sole karyotypic abnormality where it corresponded to a characteristic bone marrow picture called the "5q− syndrome." On the other hand, some classic balanced AML-associated changes – again the t(8;21), t(15;17), and inv(16) we have referred to previously, as well as translocations involving 11q23 – were rarely or never seen in MDS. It became clear from these studies that MDS was indeed a neoplastic disease, something that had not been universally accepted, and that largely the same pathogenetic chromosome aberrations in many instances lay behind both myeloid leukemias and preleukemias. Whether additional aberrations had to develop to push a myelodysplastic bone marrow clone over the brink and into full AML mode was uncertain, but obviously not always necessary, at least not if we restrict ourselves to commenting on what is cytogenetically visible. Likewise, whether AML patients with t(15;17), inv(16), and t(8;21) also pass through a short MDS phase, but one too brief to be detected, remained a possibility but again, not one supported by cytogenetic evidence. As always, many a strange hypothesis can be construed if one dares to invoke the unseen, but it is rarely scientifically sound to do so. And yet, absence of evidence is not necessarily evidence of absence.

It has been long known that AML is a hazard not only after therapeutic irradiation and/or treatment with cytostatic medications, but also after occupational exposure to various toxic compounds, including polyaromatic hydrocarbons such as benzene. Such secondary acute myeloid leukemia (sAML) tends to be particularly aggressive and difficult to treat. When large numbers of AML cases were examined in the 1970s and 1980s and the results scientifically processed, it turned out that the aberration profiles of secondary leukemias differed to some extent from that of primary AML. Patients who had been exposed to alkylating agents, radiation or industrial carcinogens often had complex, sometimes hypodiploid karyotypes in which −5, −7, and addition of unknown material to 12p and 17p were prominent features. On the other hand, those with leukemias secondary to anticancer treatment with topoisomerase II inhibitors mostly had simple karyotypes with rearrangement of chromosome band 11q23 (often due to targeting of the *KMT2A* [previously *MLL*] oncogene; see also below) as a recurring theme.

The cytogenetics of sAML thus offers insight into the age-old etiological question – for the most part studiously ignored in this book – of whether leukemogenic cytogenetic abnormalities are the product of some specific action by environmental leukemogens at particular genomic sites. The alternative would be that the induction of cancer-associated aberrations is part of an entirely stochastic process which leaves behind for analysis and detection only those cells that were transformed and multiply as a leukemic clone because of the genetic injury they suffered, while all the other random changes remained unnoticed. The latter is the null hypothesis favored by whatever conventional wisdom may be said to exist in the field, but the sAML data cannot be swept under the carpet, at least not with ease: Exposure to some toxins does give rise to specific aberration patterns in leukemic cells.

The more chronic myeloproliferative disorders (myelofibrosis, polycythemia vera, and others) also carry acquired aberrations of the "myeloid" type, making cytogenetic differentiation between them and MDS/AML impossible in many cases. In particular, −7, +8, and 20q− are such typical, but not entirely specific, abnormalities. Again, we face a situation where we do not know whether some toxins cause these particular genomic changes with exceptional ease, or whether susceptible myeloid stem cells respond with leukemic transformation to exactly these changes more readily than do other cell types.

Also two-thirds of *acute lymphocytic leukemia* (ALL) cases show clonal chromosome abnormalities at diagnosis, but mostly of another type than those found in AML and other myeloid neoplasias. Changes such as 5q−, 7q−, and 20q− are rarely if ever seen. Balanced translocations, at least those detectable by banding analysis, seem to play a smaller overall role than they do in myeloid leukemia, with the Philadelphia-producing t(9;22) and rearrangements of 11q23 as notable exceptions. The latter two also show peculiar age distributions which have not been fully accounted for: Whereas t(9;22) is clearly more common in old ALL patients, changes such as t(4;11)(q21;q23) and t(11;19)(q23;p13) (the most frequent ALL-associated 11q23- rearrangements) are more common in the young, especially children below 1 year of age. Other karyotypic features that are more characteristic of preschool than adult ALL are the occurrence of a submicroscopic translocation, t(12;21)(p13;q22) (detectable only by fluorescence *in situ* hybridization [FISH] or other molecular-genetic methods; see Chapter 11) and massive hyperdiploidy with multiple chromosomal trisomies (typically gains of 4, 6, 10, 14, 17, 18, 21, and X; double gains of 21 resulting in tetrasomy are also common) but few, if any, structural abnormalities. This being said, it sometimes happens that children suffer from the adult form of the disease, genetically speaking, and vice versa.

In lymphocytic malignancies, especially those that are more mature than ALL but also occasionally when the cells are poorly differentiated, genetic differences corresponding to the important dichotomy of B-lineage versus T-lineage differentiation exist in line with the changes immature lymphatic cells go through to become constituents of either the humoral or cellular immune system. B-precursor cells during their normal development undergo such physiological rearrangements at their immune-chain gene loci (the heavy-chain *IGH* in 14q32 and the light-chain *IGK* in 2p12 and *IGL* in 22q11), whereas T-precursor cells undergo parallel changes of their T-cell receptor genes on chromosomes 14 (alpha and delta) and 7 (beta and gamma). Probably as a reflection of these developmental propensities or retained proneness to recombine at the said loci, rearrangements targeting *IGH* (often) or the light-chain loci (less often) are found in diseases such as B-differentiated chronic lymphocytic leukemia, multiple myeloma, and non-Hodgkin lymphoma with a variety of oncogenes (see below) whose loci are found in translocation partner breakpoints.

Several of these leukemia- or lymphoma-specific translocations show considerable cytogenetic-phenotypic specificity beyond the fact that they occur in malignant cells that have begun B-lineage differentiation. The most distinct example in this regard is the t(8;14)(q24;q32) found in 75% of all Burkitt lymphomas (the variant recombinations of 2p12 [*IGK*] and 22q11 [*IGL*] make up the remaining 25%), but t(14;18)(q32;q21) in follicular non-Hodgkin lymphomas and t(3;14)(q27;q32) in diffuse large B-cell lymphomas illustrate the same picture of specificity. Also several other acquired chromosomal changes are clearly nonrandom: Trisomy 12 is a common aberration in chronic lymphocytic leukemia but may occur also in other leukemic malignancies, while 6q− is a rather unspecific lymphatic malignancy marker, sort of similar to +8 for myeloid malignancies in this regard.

The identification of specific cytogenetic aberrations in the 1970s enabled meaningful clinical-cytogenetic association studies, the most important of which were the International Workshops on Chromosomes in Leukemia. They provided an arena for fruitful, at the time unique, collaboration among cytogeneticists, clinicians, and pathologists who shared their data and insights in order to find diagnostically and prognostically interesting associations between cytogenetic aberrations and clinical characteristics in the various hematological malignancies. The results obtained over a 10-year period showed that cytogenetics could subdivide otherwise identical leukemias and lymphomas into distinct subgroups on the basis of whether specific abnormalities were present, a classification that proved to have important clinical implications. For example, the workshop collaborators demonstrated that the diagnostic karyotype in childhood acute lymphoblastic leukemia was of greater prognostic importance than any hitherto known risk factor such as patient age, white blood cell count or immunophenotype (Third International Workshop on Chromosomes in Leukemia 1983; Bloomfield et al. 1986). The studies performed on groups of patients from different parts of the world – well characterized cytogenetically and hematologically as well as clinically – were thus instrumental in consolidating cytogenetics as a well-nigh indispensable requirement for optimal care of leukemia patients.

One reason why cancer cytogenetics made such an impact not only clinically (diagnostically, prognostically, and as a means to guide therapeutic decisions; see Chapter 13), but also purely scientifically during this period was that other techniques of a more molecular nature simultaneously became available that helped explain what the observed chromosomal changes might mean in terms of cellular pathophysiology. This is a theme we will return to more extensively in Chapter 10, but it seems prudent in the context of leukemia cytogenetics to say something about one of the key concepts that changed thinking about leukemogenesis – and tumorigenesis in general – at about this time: The role of dominantly acting cancer genes or oncogenes. Given the angle of this story, we shall restrain ourselves to comments that address the changes oncogenes undergo as a consequence of cancer-specific chromosome rearrangements.

In both this and the preceding chapter, the translocations t(8;14), t(2;8), and t(8;22) have been mentioned as the cytogenetic hallmarks of Burkitt lymphoma. Rare cases of Burkitt-like leukemia also exist with genetic features identical to those of the corresponding lymphomatous disease. Molecular genetic studies have shown that the 8q24 target uniting all three rearrangements is the *MYC* oncogene locus, whereas the breaks in 14q32, 2p12, and 22q11 occur in regulatory elements whose normal function it is to stimulate transcription of the immunoglobulin genes *IGH*, *IGK*, and *IGL* in immunocompetent cells. Either the *MYC* gene moves into the vicinity of the regulatory elements controlling transcription of the immunoglobulin genes, the situation in t(8;14) (Figure 8.1), or *MYC* remains on der(8)t(2;8) or der(8)t(8;22) but the light-chain controlling elements move from 2 or 22 to der(8). The principal outcome is the same in both scenarios: No fusion gene is created by the shuffling around of genetic material nor is any fusion protein produced as the outcome of the translocation, but instead one sees increased and untimely production of MYC. The primary structure of this oncoprotein need not be abnormal in any way, which is why we refer to this type of oncogene activation as *quantitative*.

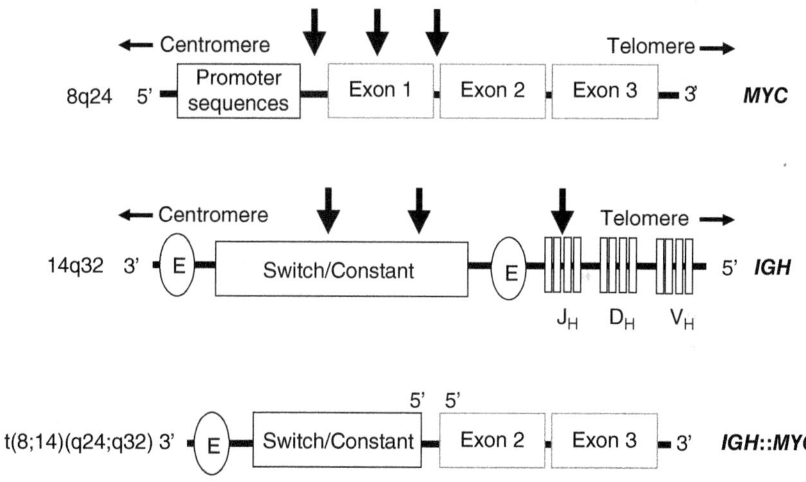

FIGURE 8.1 Gene fusion leading to upregulation. The t(8;14)(q24;q32) translocation, the most common cytogenetic rearrangement in Burkitt lymphoma, leads to deregulation of the *MYC* gene at 8q24 through its juxtaposition with regulatory elements of the immunoglobulin heavy-chain gene (*IGH*) at 14q32, i.e., *MYC* becomes constitutively activated because its expression is driven by immunoglobulin enhancers (E). The *MYC* gene has three exons and is oriented with its 5′ end toward the centromere. The breaks (arrows) vary and may be scattered over an area larger than 200 kb at the 5′ part of the gene upstream of exon 2, the first coding exon. As a consequence, the two protein-encoding exons are always spared and are translocated to the *IGH* locus in 14q32. The breaks in the *IGH* gene usually take place within switch regions but can also involve joining regions or, occasionally, a variable or constant region. *IGH* is oriented with its 5′ part toward the telomere, so the translocation leads to a 5′–5′ (head-to-head) fusion of *MYC* with sequences from the *IGH* locus. Source: Adapted from Mitelman et al. (2007).

Indeed, quantitative activation of oncogenes is *the* recurring theme for many (all?) translocations in lymphatic malignancies in which immunoglobulin or T-cell receptor genes are recombined with loci harboring protooncogenes. The differences largely hinge on which these fusion partners are. This, again, corresponds to a huge body of knowledge we shall not even try to tap into more than superficially. However, there is one lymphoma-specific translocation we *would* like to call special attention to, namely t(14;18)(q32;q21) that is the primary, sometimes sole, chromosome abnormality in many follicular non-Hodgkin lymphomas. Very occasionally, its variants t(2;18)(p12;q21) and t(18;22)(q21;q11) are seen, bringing the situation in line with findings in Burkitt lymphoma. The t(14;18), t(2;18), and t(18;22) all target a gene called *BCL2* in 18q21 leading to overproduction of an otherwise normal BCL2 protein.

The truly revolutionary take-home message from follicular lymphoma cytogenetic studies, however, the circumstance that sets this situation pathogenetically apart from Burkitt-like disease, came from the discovery that the BCL2 protein (gene names should be written in italics whereas protein names are not, but as is much too often the case with orthographic rules, this one is frequently sinned against) binds to the inner mitochondrial membrane and blocks preprogrammed cell death (apoptosis). Thus, evidence was for the first time produced, at the highest possible resolution level, that malignant accumulation of cells can result from inhibited cell death, not just uncontrolled cell division.

The majority of leukemias and lymphomas provide evidence of the other main chromosomal mechanism of protooncogene activation, namely the melting together in translocation breakpoints of genes leading to a new chimera (Figure 8.2). Again, the iconic t(9;22)(q34;q11) that gives rise to the Philadelphia chromosome in chronic myeloid leukemia (CML) and some acute leukemias is the best-known example.

Detailed cytogenetic followed by molecular genetic studies during the 1970s and 1980s showed that whereas 9q34 breakpoints target the *ABL* (now designated *ABL1*) oncogene but are relatively widespread (although of course not so widely that the differences can be seen microscopically), the breakpoints in 22q11 are much more tightly distributed, corresponding to what was first called the breakpoint cluster region and later the *BCR* gene. As a consequence of the t(9;22), a qualitatively new *BCR::ABL1* fusion gene is generated which, in its turn, produces a qualitatively new BCR::ABL1 oncoprotein that turned out to be a deregulated tyrosine kinase with elevated activity. Much more about the molecular genetic consequences of t(9;22) is of course now known, including similarities and differences between the

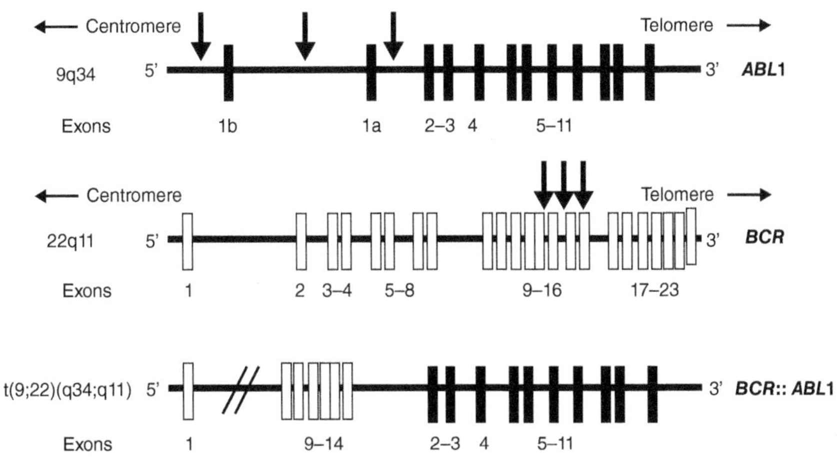

FIGURE 8.2 Gene fusion leading to a chimeric gene. The Philadelphia chromosome, which originates through the translocation t(9;22)(q34;q11), juxtaposes the 5′ part of the *BCR* gene at 22q11 with the 3′ end of the *ABL1* gene at 9q34, resulting in the creation of a hybrid *BCR::ABL1* fusion gene. The *ABL1* gene is oriented with its 5′ end toward the centromere of chromosome 9. The gene spans more than 230 kb and contains two alternative first exons, 1b and 1a, followed by exons 2–11. Exon 1b is located approximately 200 kb upstream of exon 1a. The breakpoints (arrows) are scattered over a large area (greater than 300 kb) at the 5′ end of the gene, either upstream of the first alternative exon 1b, between the two alternative exons, or between exons 1a and 2. Irrespective of the breakpoint location, splicing of the hybrid transcript yields an mRNA in which *BCR* sequences are fused to *ABL1* exon 2. The *BCR* gene has its 5′ end toward the centromere of chromosome 22, spans 135 kb, and has 23 exons. In most patients with chronic myeloid leukemia and at least one-third of patients with Philadelphia chromosome-positive acute lymphoblastic leukemia, the break occurs in a 5.8 kb major breakpoint cluster region that spans exons 12–16. Source: Adapted from Mitelman et al. (2007).

fusion genes produced in acute and chronic Ph-positive leukemias, but that is beyond the scope of our present narrative. Suffice it to emphasize that the Ph chromosome is the site of a fusion gene, a qualitatively altered gene that does not exist outside the neoplastic context. It produces a correspondingly altered oncoprotein that can serve as a target for therapeutic intervention (see Chapter 13). The hunt for new oncogenic fusion genes has since been at the forefront of scientific cytogenetic analyses of malignant cells.

When studies using the polymerase chain reaction (PCR) technique detected traces of *BCR::ABL1* fusion genes in some nonleukemic, healthy individuals, admittedly in 1000- or 10,000-fold smaller concentrations than in patients with Ph-positive leukemia, it created considerable consternation. As we see it, the most likely explanation for the situation is that the gene fusion may very occasionally also exist in cells that cannot respond by leukemic transformation; the seed may be there but the soil is not right. The situation therefore does not constitute such a conundrum as one might at first think, but it does warn that one should be careful about searching for disease-specific changes at a resolution (or concentration) level higher (lower) than it should be. The use of oversensitive techniques can lead to false positives and a lot of clinical trouble.

Whereas many cancer-specific translocations lead to fusions of only two genes with no or few variations on the theme, acute leukemia cytogenetics is full of examples of the opposite, namely that a "promiscuous" gene or breakpoint can recombine with a lot of translocation partners to generate partly similar fusions or other gene alterations that are all leukemogenic. In these situations, the pathogenic outcome is the final common path of many more or less similar variations played out on the same theme, and we still know too little about what the clinical consequences might be of this "more or less" variability. Prominent examples of such highly reactive leukemia-associated genes are *ETV6* in 12p13 and *RUNX1* in 21q22. However, the most "promiscuous" oncogene of them all is *KMT2A* (previously called *MLL*) in 11q23 which is known to recombine in both lymphoblastic and myeloblastic leukemias, i.e., ALL and AML, with well above 100 gene partners corresponding to almost the same number of chromosomal translocations and a few other intra- and interchromosomal rearrangements. Some of these recombinations, the most common being t(4;11)(q21;q23), typically lead to leukemias that are called ALL, whereas others, with t(9;11)(p21;q23) as the most prominent example, give rise to AML. However, the literature also contains cases of t(9;11)-positive leukemias that were diagnosed as ALL and t(4;11)-positive cases called AML.

To what extent does this variability reflect significant reality? Should a leukemia primarily be "defined" – and hence diagnosed and treated – by the pattern of leukemogenic changes that gave rise to it, or should it, as has traditionally been the case, be named after the phenotypic features of the leukemic cells? Practice is gradually changing with regard to this very principled question in the direction of paying more attention to pathogenesis than cellular morphology. Thus, genetic features are increasingly being incorporated as necessary parameters in the WHO classifications of both leukemias and lymphomas (Swerdlow et al. 2008; Arber et al. 2016). A similar development is also seen elsewhere in oncology as a result of advances made in solid tumor cytogenetics, to be covered in the next chapter.

REFERENCES AND FURTHER READING

Arber, D.A., Orazi, A., Hasserjian, R. et al. (2016). The 2016 revision to the World Health Organization classification of myeloid neoplasms and acute leukemia. *Blood* 127: 2391–2405.

Bennett, J.M., Catovsky, D., Daniel, M.T. et al. (1976). Proposals for the classification of the acute leukemias. French-American-British (FAB) co-operative group. *Br. J. Haematol.* 33: 451–458.

Bloomfield, C.D., Goldman, A.I., Alimena, G. et al. (1986). Chromosomal abnormalities identify high-risk and low-risk patients with acute lymphoblastic leukemia. *Blood* 67: 415–420.

Heim, S. and Mitelman, F. (eds.) (2015). *Cancer Cytogenetics. Chromosomal and Molecular Genetic Aberrations of Tumor Cells*, 4e. Oxford: Wiley-Blackwell.

Mitelman, F., Johansson, B., and Mertens, F. (2007). The impact of translocations and gene fusions on cancer causation. *Nat. Rev. Cancer* 7: 233–245.

Rowley, J.D., Le Beau, M.M., and Rabbitts, T.H. (eds.) (2015). *Chromosomal Translocations and Genome Rearrangements in Cancer*. New York: Springer.

Swerdlow, S.H., Campo, E., Lee Harris, N. et al. (eds.) (2008). *WHO Classification of Tumours of Haematopoietic and Lymphoid Tissues*. Lyon: IARC.

Third International Workshop on Chromosomes in Leukemia (1983). Chromosomal abnormalities and their clinical significance in acute lymphoblastic leukemia. *Cancer Res.* 43: 868–873.

Tosi, S. and Reid, A.G. (eds.) (2016). *The Genetic Basis of Haematological Cancers*. Chichester: John Wiley & Sons.

CHAPTER 9

Solid Tumor Cytogenetics

Throughout the twentieth century, while chromosome examinations of hematopoietic neoplastic cells were taking step after uneven step forward both technologically and conceptually, analyses of solid cancers always lagged far behind. When the body of cytogenetic knowledge about leukemias and lymphomas finally expanded rapidly during the two first decades following the introduction of chromosome banding, the 1970s and 1980s, many workers in the field wondered what might be the differences and similarities between the aberrations thus detected and those of solid tumor cells. Would the main lessons learned from studies of neoplastic bone marrows and lymph nodes prove relevant also for carcinomas, sarcomas, and other solid cancers, or would novel findings be made that necessitated alternative ways of looking at how chromosomal changes contribute to neoplastic transformation of previously normal somatic cells? What kind of tissue or cell type specificities were to be expected? Many interesting possibilities were suggested, and in hindsight it is perhaps surprising that more effort was not put into solid tumor cytogenetics than was the case during the 1970s.

When several cancer cytogenetic research groups after 1980 finally decided to complement their studies of hematological malignancies with comparable examinations of solid tumors, swift progress was made. Table 9.1 summarizes 24 major discoveries in solid tumor cytogenetics made during that decade, a truly impressive number. It is important to bear in mind, however, that some relevant knowledge had been obtained even before banding techniques became available. (i) Meningioma cells were known to be frequently monosomic for a G-group chromosome, often as the only

Abnormal Chromosomes: The Past, Present, and Future of Cancer Cytogenetics.
Sverre Heim and Felix Mitelman.

TABLE 9.1 Characteristic cytogenetic aberrations detected by banding analyses of solid tumors 1980–1989.

Year	Tumor type	Aberration	References
1980	Salivary gland adenoma	t(3;8)(p21;q12)	Mark et al. (1980)
1982	Germ cell tumors	i(12)(p10)	Atkin and Baker (1982)
	Lung cancer	del(3)(p23p14)	Whang-Peng et al. (1982)
	Retinoblastoma	i(6)(p10)/del(13q)	Balaban et al. (1982), Kusnetsova et al. (1982)
	Rhabdomyosarcoma (alveolar)	t(2;13)(q36;q14)	Seidal et al. (1982)
1983	Ewing sarcoma	t(11;22)(q24;q12)	Aurias et al. (1983), Turc-Carel et al. (1983)
	Salivary gland adenoma	der(12)(q13–15)	Stenman and Mark (1983)
	Wilms' tumor	der(16)t(1;16)(q21;q13)	Kaneko et al. (1983)
1985	Chondrosarcoma (myxoid)	t(9;22)(q31;q12)	Hinrichs et al. (1985)
1986	Kidney cancer	t(X;1)(p11;q21)	de Jong et al. (1986)
	Lipoma	t(3;12)(q27;q13)	Heim et al. (1986), Turc-Carel et al. (1986)
	Liposarcoma (myxoid)	t(12;16)(q13;p11)	Limon et al. (1986a)
	Synovial sarcoma	t(X;18)(p11;q11)	Limon et al. (1986b)
1987	Kidney cancer	del(3p)/der(3)t(3;5)(p13;q22)	Kovacs et al. (1987)
	Liposarcoma (highly differentiated)/atypical lipomatous tumor	Ring chromosome(s)	Heim et al. (1987)
	Lipoma	der(12)(q13–15)	Mandahl et al. (1987)
1988	Primitive neuroectodermal tumor	i(17)(q10)	Griffin et al. (1988)

(continued)

TABLE 9.1 *(continued)*

Year	Tumor type	Aberration	References
	Salivary gland cystadenolymphoma	t(11;19)(q21;p13)	Bullerdiek et al. (1988)
	Uterine leiomyoma	del(7)(q22q31)	Boghosian et al. (1988)
	Uterine leiomyoma	t(12;14)(q14;q24)	Heim et al. (1988), Mark et al. (1988), Turc-Carel et al. (1988)
1989	Infantile fibrosarcoma	+8,+11,+20	Mandahl et al. (1989), Speleman et al. (1989)
	Lipoma	der(6)(p21)	Sait et al. (1989)
	Ovarian cancer	add(19)(p13)	Pejovic et al. (1989)

aberration. (ii) Abnormal karyotypes with clearly nonrandom aberration patterns had been demonstrated in experimentally induced sarcomas in different animal species. (iii) Ascites fluid tumor cells from patients with metastatic cancers were known to be highly aneuploid in addition to carrying multiple structural rearrangements. Beyond any doubt, therefore, many solid tumors carried acquired chromosome abnormalities, so at least as far as they were concerned overall similarity existed with the leukemia situation.

A major general difference between solid tumors on the one hand and bone marrow and lymph node neoplasms on the other is that no truly benign counterpart exists among the latter to the many nonmalignant epithelial and connective tissue tumors that form such a significant part of human pathology. Perhaps this merely reflects the fact that bone marrow malignancies do not develop within the confines of a fixed and rigid stromal structure – solid tumors are called exactly that, solid, for a reason – so possibly we are in hematology dealing with more or less floating disease processes from the very beginning. Nevertheless, could it be that some distinctive chromosomal characteristics are typical of benign tumors exclusively, setting them unequivocally apart from malignant ones?

If we stop and think for a moment, this does not seem to be a very likely proposition, at least not if one tries to view the entire disease process from the perspective of the neoplastic cells, not the host organism or patient. The most significant leap in the development of a neoplasia is the transformation event itself, which we, who follow in the footsteps of Boveri, believe is equivalent to the acquisition of a primary chromosome abnormality characteristic of that particular disease. Whether the neoplasm is able to destroy surrounding tissue and set up distant metastases, the two distinguishing features of malignancies, comes across as less important from the transformed cell's point of view. Indeed, the destruction of a host organism leading to its death is, evolutionarily speaking, a thoroughly unsuccessful strategy since this would inevitably lead to the neoplastic clone's extinction as well. However crucial the words *benign* and *malignant* may be clinically, they are irrelevant from the standpoint of neoplastic cells. According to this line of reasoning, one would not expect benign tumors to harbor fundamentally different chromosome changes compared to malignant ones, although some systematic differences might exist accounting for the variability in clinical impact of the two situations.

When we in the mid-1980s set out to study two very common and completely innocuous benign connective tissue tumors – lipomas and leiomyomas – the results obtained soon surpassed expectations by a very wide margin. Thinking back on it, it is curious to remember that some collaborating surgeons and pathologists were rather disenchanted when learning that attempts would be made to find out whether these tumors carried chromosomal aberrations, for never are mitotic figures seen in lipoma or leiomyoma sections nor do these tumors transform malignantly even after decades of observation. Would it even be possible to culture *in vitro* the relevant cells from such neoplasms?

It turned out to be possible indeed, although probably the cells that attach and begin dividing are not the highly differentiated ones that dominate the microscopic picture, but rather similar-looking cells that have retained more "primitive" features, including the ability to enter mitosis. Not only could short-term cultures be established from them with unexpected ease, but they also yielded technically excellent chromosome preparations.

The findings were rich, but not too rich; the aberration pattern unfolding was anything but chaotic. Clear-cut specificities were detected for both lipomas and leiomyomas, although for each tumor entity it soon became clear that several pathogenetic pathways could lead to the same phenotypic endpoint. A feature common to many tumors of both types was rearrangement of the chromosomal region 12q13-15, albeit with different predominant translocation partners in fat cell and smooth muscle tumors.

Around two-thirds of the lipomas studied cytogenetically turned out to have rearrangements of 12q13-15, with t(3;12)(q27-29;q13-15) as by far the most common translocation (Figure 9.1). Other relatively frequent partner breakpoints mapped to 1p32-34, 2p22-24, 2q35-37, 5q32-34, 12p11-12, and 12q24, and then there were many more rare alternatives. Less common

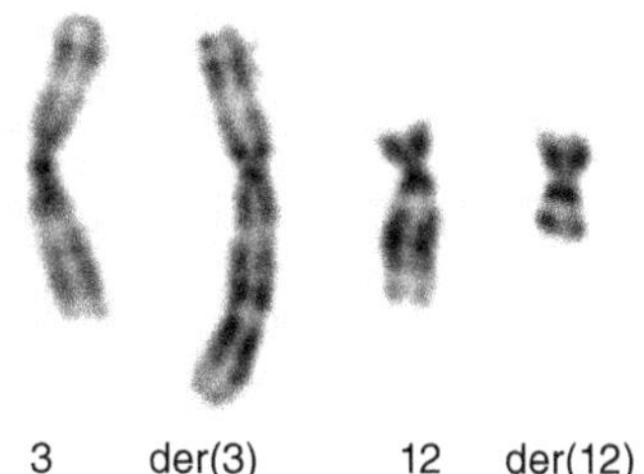

FIGURE 9.1 Translocation t(3;12)(q28;q14) in a lipoma. Source: Mandahl and Mertens (2015).

cytogenetic subgroups that do not have any visible involvement of 12q also exist, some of which are defined by deletion of 13q, translocations involving 6p21-23, or rearrangement of proximal 8q (8q11-13) (Mandahl and Mertens 2015).

Pathologists would not be pathologists if they did not have a keen eye for small phenotypic differences even among benign tumors showing fat cell differentiation – which is the defining feature of lipomas – and so a number of different tumor subgroups are known that show partial covariation with the above-mentioned or other genotypic features. Many chondroid lipomas carry an 11q13;16p13-translocation, 8q11-13 rearrangements are typical of lipoblastomas, spindle cell lipoma and pleomorphic lipoma often have 13q-losses and so on almost *ad infinitum*. How meaningful it is to split up diagnostic entities in this manner is a question of immense importance but one not easily answered; we shall return to it repeatedly. It seems to be one of the many inexplicable facts of life that, in whatever context, some research-inclined individuals tend to be splitters whereas others are lumpers.

Benign smooth muscle tumors share with lipomas a propensity for 12q13-15 rearrangements, but in these tumors the translocation t(12;14)(q14-15;q23-24) is quantitatively dominant (Figure 9.2). Other common

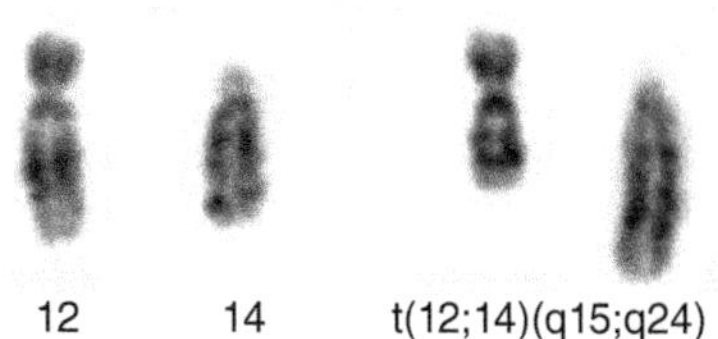

FIGURE 9.2 Translocation t(12;14)(q15;q24) in a uterine leiomyoma. Source: Micci and Heim (2015).

cytogenetic leiomyoma subgroups are defined by a very distinct deletion of 7q, del(7q21.2q31.2), trisomy 12, rearrangements involving 6p21, 10q, and 1p, deletions of 3q, and changes of the X chromosome. For many of these aberrations, we still do not know how they contribute to neoplastic transformation; this is especially so for the deletions, not to mention the numerical gains, in this context trisomy 12 (Micci and Heim 2015). But for some of the most characteristic changes we now do have a fair understanding of their pathogenetic impact, thanks to the happy marriage between cytogenetics and molecular genetics referred to in the previous chapter, a theme that will recur in the remainder of this book.

In both lipomas and leiomyomas, the rearrangements targeting 12q14 and its immediate surroundings (the variable assignments to bands 12q13-15 in the relevant literature more likely reflect the imprecision of cytogenetic techniques when studying the very small than anything else) act through alteration of the high-mobility group AT-hook 2 (*HMGA2*) gene situated there. The breakpoints differ considerably at the molecular level although many of them are inside the gene locus, often in the large third intron, seeing to it that a fusion gene is generated between *HMGA2* and the translocation partner, mostly leading to ectopic sequences replacing the acidic C-terminus of HMGA2. On many other occasions, however, the break takes place outside the gene itself. So when it comes to *HMGA2* involvement in lipomas, and this has been shown to apply even to some tumors that have no cytogenetically visible rearrangement of 12q14, the situation turns out to be extremely intricate. Sometimes a genuine fusion gene is generated, sometimes regulatory changes take place, and on yet other occasions the gene and its product are truncated via mechanisms that are not yet fully understood. But whenever this happens in a stem cell already set on proceeding down a pathway of terminal fat cell differentiation, neoplastic transformation occurs and a lipoma is the end-result.

In other situations, presumably when the transformed somatic cell is bound to differentiate in the smooth muscle direction, the clone grows into a leiomyoma. When this is the case, the most common partner gene comes from 14q (typically *RAD51L1*) rather than 3q (*LPP*), but also a wide array of other cytogenetic as well as molecular genetic changes and mechanisms may come into play to alter the *HMGA2* gene and/or deregulate its transcription. Indeed, breakpoints outside the latter locus are more common than is the formation of actual fusion genes, but again, untimely production of an HMGA2 protein that may or may not have an altered primary structure is the key factor in establishing the neoplasm. In fact, whereas generation of a fusion gene between *HMGA2* and one of many partners seems to be the general rule in lipomas, leiomyomas more often arise through a strictly deregulatory route (Quade et al. 2003). The pathogenetic overlap is considerable between the two types of benign connective tissue tumors, however.

The above description of the role of *HMGA2* alteration and deregulation in neoplasia is by no means complete – it cannot be otherwise given that our focus is on the chromosomal changes in cancer and related diseases, not the molecular consequences they lead to – but we would like to mention two additional points before we move on. First, another member of the HMGA family of genes, *HMGA1*, is located in band 6p21 which is one of the chromosomal hotspots for rearrangements in lipomas and leiomyomas as well as other benign connective tissue tumors. This gene is the target of 6p21 rearrangements in the said tumors. It thus seems that deregulatory alterations of *HMGA1* and *HMGA2*, obtained via a variety of chromosomal mechanisms, can have the same tumorigenic effect on suitably primed target cells. Second, deregulation of *HMGA2* not only occurs in benign connective tissue tumors but has also, albeit rarely, been found in myelodysplasia/acute myeloid leukemia (Nyquist et al. 2012). We can only conclude that the interplay between genetic and other biological variables in both benign and malignant neoplastic disorders is incredibly rich, even if we restrict our view to but a small sector of the phenotypic totality and the acquired genetic changes that characterize these neoplasms. It seems a safe bet that the store of surprises attached to the story of t(3;12), t(12;14), and *HMGA* deregulation is still far from exhausted. Little did we know that so much excitement would follow when we began studying the cytogenetics of benign fat and smooth muscle cell tumors 35 years ago.

Are there any genetic similarities between benign and malignant connective tissue tumors? The short and unsophisticated answer is no, which perhaps makes good sense since the former are not prone to develop into the latter regardless of how long they are observed.

Malignant smooth muscle tumors, leiomyosarcomas, usually have very complex karyotypes without otherwise distinguishing cytogenetic features, especially not anything that resembles the chromosomal characteristics known from leiomyoma studies. Another type of uterine tumor (the vast majority of cytogenetically studied leiomyomas and leiomyosarcomas were of uterine origin) that sometimes may represent a difficult differential diagnosis to (cellular) leiomyoma or leiomyosarcoma is endometrial stromal tumor or sarcoma. These tumors develop from stromal elements of the endometrium, hence their name, and may show different phenotypes ranging from the completely benign to clearly malignant. Although they do display highly characteristic cytogenetic as well as molecular genetic

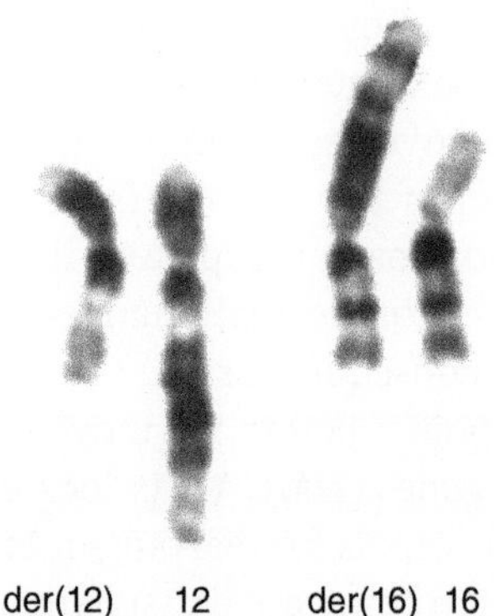

FIGURE 9.3 Translocation t(12;16)(q13;p11) characterizing myxoid liposarcoma. Source: Mandahl and Mertens (2015).

abnormality profiles (Micci et al. 2016), there is no overlap with leiomyomas and leiomyosarcomas.

Malignant fat cell tumors, liposarcomas, were traditionally divided into four different entities based on histological appearance and clinical behavior: Highly differentiated liposarcomas (HDLS), myxoid liposarcomas, round cell liposarcomas, and pleomorphic liposarcomas. The latter are the least differentiated and most malignant. Their karyotypic profile is typical of many highly malignant tumors; numerous and complex aberrations without any truly specific rearrangements is the rule. Round cell and myxoid liposarcomas may have phenotypic differences corresponding to their names, but cytogenetic studies have shown that they are karyotypically similar inasmuch as both tumor types almost always have t(12;16)(q13;p11) or a variant thereof as the primary change (Figure 9.3). Efforts have been made to identify genomic features that show covariation with the phenotypic differences between round cell and myxoid tumors and some such changes were detected (Demicco et al. 2012), but evidently these two liposarcoma subgroups start off acquiring the same primary abnormality, something that makes them pathogenetically very similar or identical.

A final comment about cytogenetic/phenotypic similarities/differences is due to what most pathologists call highly (or well) differentiated liposarcomas (HDLS), but which in the language of others go by names such as atypical lipomatous tumors (ALT). If not successfully removed, these may evolve into dedifferentiated liposarcomas, especially if the tumor is located retroperitoneally. Though some diagnosticians claim that clear phenotypic differences exist among these entities, even between HDLS and ALT, the fact remains that all three are cytogenetically indistinguishable. Supernumerary rings and/or giant marker chromosomes are their defining karyotypic features reflecting a characteristic pattern of genomic amplification that always involves 12q, often 1q material, but occasionally material from other genomic sites, too. The nomenclature confusion is due not only to naming traditions and whether subtle morphological differences can or cannot be seen upon microscopic inspection of tumor sections, but also to the fact that many of them are on the borderline between what is clearly malignant and what is not. However crucial this divide is from a clinical point of view, it is not always easy to uphold. Regardless, patients are indebted to the many extremely competent surgeons who are able to effect lasting cure whenever tumors are radically removable, almost irrespective of their differentiation; good surgery often does that no matter which names pathologists use for the disease condition.

It is far from our aim in this book to cover the totality of solid tumor cytogenetics, just as it was not our aim to discuss or even list all characteristic chromosome aberrations found in leukemic and lymphomatous diseases; the reader is referred to reviews and textbooks, such as Sandberg and Meloni-Ehrig (2010) and Heim and Mitelman (2015), for that. We chose to dwell on the above-mentioned connective tissue tumors because they illustrate some classic problems in disease definition and classification, but also because they illustrate an overriding pathogenetic similarity with hematological malignancies. In both fields, many (but not all) disease entities are characterized by distinctive cytogenetic aberration patterns. Sometimes we even find disease-specific chromosomal rearrangements in the tumor parenchyma cells, and in many of these situations molecular geneticists have taught us which pathogenetic gene-level changes the microscopically visible aberrations correspond to.

That pathogenetic similarity between (many) leukemias and (many) sarcomas exists can also be inferred from examples outside the realms of fat cell- and smooth muscle-differentiated tumors. Alveolar rhabdomyosarcoma often has a t(2;13)(q36;q14), Ewing sarcoma has a t(11;22)(q24;q12), synovial sarcoma a t(X;18)(p11;q11), clear cell sarcoma of soft tissues a t(12;22)(q13;q12), and so on. For all these sarcoma-specific chromosomal rearrangements, the corresponding pathogenetically important fusion genes have been identified, contributing mightily to an understanding of the tumorigenic process, as they also have for the various variant translocations that exist recombining the crucial oncogene with another partner. But at the same time, just as many leukemias do not seem to carry any characteristic translocation and/or fusion gene but instead have a more complex karyotype including many numerical aberrations, so also some sarcoma types display an unruly, even extremely complex, chromosome aberration pattern without distinctive molecular genetic features. Among them are the leiomyosarcomas and pleomorphic liposarcomas already mentioned, but also the most numerous subgroups of osteosarcoma and chondrosarcoma.

What about the massive majority of solid cancers in humans, the malignant tumors that display epithelial differentiation features earning them the name carcinomas? To what extent do they cytogenetically resemble leukemias, lymphomas, and sarcomas? Again, we come down to answering that some similarities certainly exist, but the differences are also pronounced. In this sense, leukemias and sarcomas are closer to one another than either is similar to carcinomas.

Before we dwell on some of the distinguishing features of common carcinomas, it is worth pointing out that cytogenetic studies of such tumors have been less extensive by a large margin than studies of leukemias and even sarcomas. In fact, less than 10% of the cancer cytogenetic data available today were gathered from malignant epithelial tumors in spite of the fact that the latter are responsible for 90% of all human cancer morbidity and mortality. At first, researchers found it prohibitively difficult to study and karyotype epithelial tumors, then the advent of powerful molecular techniques took over and led many people within the field to abandon chromosome banding analyses. As a result, much remains unexplored in carcinoma cytogenetics; probably a lot lies unknown that could and should have been known by now. Absence of evidence is not evidence of absence. This saying may have general validity but it definitely holds true in scientific contexts and, more specifically, when cancer chromosomes are concerned.

The general overall impression is nevertheless strong; whereas leukemias, lymphomas, and benign as well as malignant connective tissue tumors often have fairly simple karyotypes with perhaps a single rearrangement that carries an activated oncogene, such simplicity is absent from colorectal, lung, cervical, pancreatic, ovarian, and other common carcinomas. Exceptions exist, but mostly involving rare tumor subtypes. This is not equivalent to saying that all these carcinomas are cytogenetically indistinguishable; some features are undoubtedly more common in one tumor type, others in others, but the sharp specificities often encountered in sarcoma cytogenetics, enabling precise and certain diagnoses based on which chromosome aberrations one finds, are mostly lacking. Of necessity, therefore, although one can often conclude that a given aberration pattern shows that a malignant process has been sampled, exactly which one may be impossible to determine.

FIGURE 9.4 Metaphase from a highly malignant epithelial tumor showing numerous numerical and structural abnormalities.

For example, a deletion in 3p was early on reported to be specific for small cell lung cancer (Whang-Peng et al. 1982) but was later found in many other epithelial malignancies, too, including nonsmall cell lung cancers, kidney carcinomas, and breast cancer, all of which make its value as a diagnostic marker less than was initially hoped. Likewise, carcinomas of many types as well as other malignant tumors may have complete or partial gain of 1q, loss of 6q, gain of 8q (often as an isochromosome for the long arm), and so on. In short, the majority of the common carcinomas, the main oncological killers, have complex karyotypes with many numerical and structural chromosome abnormalities. Figure 9.4 shows an example of the bewildering complexity sometimes detected in malignant epithelial tumors.

There are some exceptions to this rule of thumb, however, in particular endometrial adenocarcinomas. These tumors tend to be deceptively simple, showing only gain of 1q-material, often as a supernumerary i(1q). Their gynecological-anatomical cancerous neighbors, on the other hand, carcinomas of the uterine cervix and the tubes/ovaries, are usually very complex.

In the latter locale, benign, borderline, and clearly malignant epithelial tumors are fairly common, making a comparison possible between clinically dangerous and innocent neoplastic processes (Micci and Heim 2015). Adenomas, cystadenomas, and even borderline ovarian tumors tend to be karyotypically simple, usually with only a few or a single numerical chromosome aberration (+12 is the most common), whereas the more malignant tumors typically are cytogenetically complex. One wonders whether the few carcinomas with trisomy 12 as the sole change that exist in the cytogenetic literature, really deserved the malignant diagnosis they received. With regard to this particular trisomy, it is intriguing to note that several other benign ovarian tumors (fibromas, fibrothecomas, fibroadenomas) have been reported with the same aberration. Nobody knows why, nor does anybody know how such a gain of a whole chromosome acts pathogenetically.

Another look at the relationship between benign and malignant tumors of the same tissue can be obtained by cytogenetic comparison of colorectal carcinomas and adenomas showing few or more extensive dysplastic features; the latter have long been thought of as developmental forerunners of malignant large bowel tumors. Bardi et al. (1997) found clonal chromosome aberrations even in hyperplastic polyps, confirming their neoplastic nature, but fewer than were seen in dysplastic adenomas. In anatomically clearly distinct adenomas from the same bowel, sometimes the same structural abnormalities (in particular 1p−, another change that is typical of colorectal carcinogenesis) were found. Either spreading had occurred of a neoplastic lesion that by conventional pathological criteria was not yet malignant, a conclusion that is hard to accept, or seemingly identical chromosomal rearrangements had been acquired by two separate and independent target cells, giving rise to synchronous neoplasms. If the latter explanation is correct, then it seems tempting to suggest that an as yet unknown carcinogen may have induced the same loss of 1p material in at least two places, but of course this all remains entirely speculative.

We shall return in Chapter 12 to a more principled discussion of the clonal relationships among primary tumors, their local recurrences, and metastatic lesions, but we would like to offer a few comments here about the cytogenetic relationships that exist among the more or less synchronous lesions that are fairly common in some cancers. In addition to the brief coverage above of the cytogenetics of tumors existing simultaneously in surgically removed large intestines, we have had the opportunity to examine such situations especially in tumors of the breast and urinary bladder.

In the bladder, synchronous polyps or even malignant tumors are common. This always seemed intuitively unsurprising to us, for after all the transitional epithelium of that organ lies bathed in a sea of carcinogens excreted by the kidneys. If field cancerization (Slaughter et al. 1953) plays any role at all in common human malignancies, then the bladder situation ought to be a perfect example (perhaps in addition to cancers of the skin, lung, and intestinal epithelium); one would expect to see often the simultaneous growth of several clonally independent tumors. When such tumors were examined cytogenetically, however, in cultures from tumors that were anatomically clearly separate in the bladder mucosa (Fadl-Elmula et al. 1999), they were mostly found to harbor identical or nearly identical chromosome aberrations. The conclusion seemed inevitable that all the lesions were part of one single neoplastic process. Again, submucosal spreading of some sort must have taken place, unless the bladder mucosa cells are particularly prone to develop a small group of characteristic, tumorigenic aberrations.

While breast cancers are not as frequently multifocal as are benign and even malignant bladder tumors, examples of simultaneity are by no means rare. More than one tumor may be found in different quadrants of the same breast, or breast cancer can be bilateral. Opinions have differed as to how these situations should be interpreted; do they represent a metastatic process or are both/all tumor lesions primary? Teixeira et al. (2002) used banding cytogenetics, which is uniquely well suited to examine the clonal relationships among tumor lesions, to trace the developmental relationship among various breast cancer lesions in the same patient. We found that sometimes no clonal cytogenetic similarity was seen between or among them, but on other occasions the tumors, even if they were not within the same breast, were clearly part of the same clonal process. Not surprisingly, ipsilateral lesions were more often clonally related than were bilateral ones. Another highly unexpected finding that came out of these extensive cytogenetic studies of breast carcinomas was that seemingly unrelated clones were often detected in cultures from the same tumor. What this might mean will again be the topic of discussion in Chapter 12.

A final comment of a more general nature is warranted about the importance of finding small clones with simple numerical chromosome aberrations, especially trisomy 7 but sometimes also other low-grade trisomies, in short-term cell cultures established from solid tumors. These aberrations have been observed by many groups of researchers who used different techniques and examined a wide range of tumors, and yet nobody knows what they signify. At one point in time, the situation was so annoying that we decided to test cells cultured from other, unquestionably nonneoplastic disease lesions (we used samples from pyelonephritic kidney pelvis and brain tissue removed because of trauma) and found trisomy 7 there, too. Obviously, this particular change can occur clonally without meaning that the cells are part of any neoplastic process (Johansson et al. 1993). Some have suggested that the +7 exists in inflammatory cells accumulating in the injured tissue, but this is hardly more than a partial explanation. The small but abnormal "clones" remain a biologically intriguing phenomenon regardless of what they signify, though probably they do not all stem from one single transformed mother cell.

From the standpoint of cytogenetic diagnosis, the important take-home lesson deserves to be repeated: The finding of a "clonal" chromosome aberration ("clonal" is here used in accordance with the standard operational definition, i.e., the change is present in at least two metaphase cells) in a solid disease lesion should not automatically be translated into a conclusion that a neoplastic process is ongoing. A certain quality assessment of the findings is always necessary in solid tumor cytogenetics.

REFERENCES AND FURTHER READING

Atkin, N.B. and Baker, M.C. (1982). Specific chromosome change, i(12p), in testicular tumors? *Lancet* 2: 1349.

Aurias, A., Rimbaut, C., Buffe, D. et al. (1983). Chromosomal translocations in Ewing's sarcoma. *N. Engl. J. Med.* 309: 496–497.

Balaban, G., Gilbert, F., Nicholas, W. et al. (1982). Abnormalities of chromosome #13 in retinoblastomas from individuals with normal constitutional karyotypes. *Cancer Genet. Cytogenet.* 6: 213–221.

Bardi, G., Parada, L.A., Bomme, L. et al. (1997). Cytogenetic comparisons of synchronous carcinomas and polyps in patients with colorectal cancer. *Br. J. Cancer* 76: 765–769.

Boghosian, L., Dal Cin, P., and Sandberg, A.A. (1988). An interstitial deletion of chromosome 7 may characterize a subgroup of uterine leiomyoma. *Cancer Genet. Cytogenet.* 34: 207–208.

Bullerdiek, J., Haubrich, J., Meyer, K., and Bartnitzke, S. (1988). Translocation t(11;19)(q21;p13.1) as the sole chromosome abnormality in a cystadenolymphoma (Warthin's tumor) of the parotid gland. *Cancer Genet. Cytogenet.* 35: 129–132.

Demicco, E.G., Torres, K.E., Ghadimi, M.P. et al. (2012). Involvement of the PI3K/Akt pathway in myxoid/round cell liposarcoma. *Mod. Pathol.* 25: 212–221.

Fadl-Elmula, I., Gorunova, L., Mandahl, N. et al. (1999). Cytogenetic monoclonality in multifocal uroepithelial carcinomas: evidence of intraluminal tumour seeding. *Br. J. Cancer* 81: 6–12.

Griffin, C.A., Hawkins, A.L., Packer, R.J. et al. (1988). Chromosome abnormalities in pediatric brain tumors. *Cancer Res.* 48: 175–180.

Heim, S. and Mitelman, F. (eds.) (2015). *Cancer Cytogenetics. Chromosomal and Molecular Genetic Aberrations of Tumor Cells.* Chichester: Wiley-Blackwell.

Heim, S., Mandahl, N., Kristoffersson, U. et al. (1986). Reciprocal translocation t(3;12)(q27;q13) in lipoma. *Cancer Genet. Cytogenet.* 23: 301–304.

Heim, S., Mandahl, N., Kristoffersson, U. et al. (1987). Marker ring chromosome – a new cytogenetic abnormality characterizing lipogenic tumors? *Cancer Genet. Cytogenet.* 24: 319–326.

Heim, S., Nilbert, M., Vanni, R. et al. (1988). A specific translocation, t(12;14)(q14-15;q23-24), characterizes a subgroup of uterine leiomyomas. *Cancer Genet. Cytogenet.* 32: 13–17.

Hinrichs, S.H., Jaramillo, M.A., Gumerlock, P.H. et al. (1985). Myxoid chondrosarcoma with a translocation involving chromosomes 9 and 22. *Cancer Genet. Cytogenet.* 14: 219–226.

Johansson, B., Heim, S., Mandahl, N. et al. (1993). Trisomy 7 in nonneoplastic cells. *Genes Chromosomes Cancer* 6: 199–205.

de Jong, B., Molenaar, I.M., Leeuw, J.A. et al. (1986). Cytogenetics of a renal adenocarcinoma in a 2-year-old child. *Cancer Genet. Cytogenet.* 21: 165–169.

Kaneko, Y., Kondo, K., Rowley, J.D. et al. (1983). Further chromosome studies on Wilms' tumor cells of patients without aniridia. *Cancer Genet. Cytogenet.* 10: 191–197.

Kovacs, G., Szücs, S., de Riese, W., and Baumgärtel, H. (1987). Specific chromosome aberration in human renal cell carcinoma. *Int. J. Cancer* 40: 171–178.

Kusnetsova, L.E., Prigogina, E.L., Pogosianz, H.E., and Belkina, B.M. (1982). Similar chromosomal abnormalities in several retinoblastomas. *Hum. Genet.* 61: 201–204.

Limon, J., Turc-Carel, C., Dal Cin, P. et al. (1986a). Recurrent chromosome translocations in liposarcoma. *Cancer Genet. Cytogenet.* 22: 93–94.

Limon, J., Dal Cin, P., and Sandberg, A.A. (1986b). Translocations involving the X chromosome in solid tumors: presentation of two sarcomas with t(X;18)(q13;p11). *Cancer Genet. Cytogenet.* 23: 87–91.

Mandahl, N. and Mertens, F. (2015). Soft tissue tumors. In: *Cancer Cytogenetics. Chromosomal and Molecular Genetic Aberrations of Tumor Cells* (eds. S. Heim and F. Mitelman), 583–614. Chichester: Wiley-Blackwell.

Mandahl, N., Heim, S., Johansson, B. et al. (1987). Lipomas have characteristic structural chromosomal rearrangements of 12q13-q14. *Int. J. Cancer* 39: 685–688.

Mandahl, N., Heim, S., Rydholm, A. et al. (1989). Nonrandom numerical chromosome aberrations (+8, +11, +17, +20) in infantile fibrosarcoma. *Cancer Genet. Cytogenet.* 40: 137–139.

Mark, J., Dahlenfors, R., Ekedahl, C., and Stenman, G. (1980). The mixed salivary gland tumor – a normally benign human neoplasm frequently showing specific chromosomal abnormalities. *Cancer Genet. Cytogenet.* 2: 231–241.

Mark, J., Havel, G., Grepp, C. et al. (1988). Cytogenetical observations in human benign uterine leiomyomas. *Anticancer Res.* 8: 621–626.

Micci, F. and Heim, S. (2015). Tumors of the female genital organs. In: *Cancer Cytogenetics. Chromosomal and Molecular Genetic Aberrations of Tumor Cells* (eds. S. Heim and F. Mitelman), 519–556. Chichester: Wiley-Blackwell.

Micci, F., Gorunova, L., Agostini, A. et al. (2016). Cytogenetic and molecular profile of endometrial stromal sarcoma. *Genes Chromosomes Cancer* 55: 834–846.

Nyquist, K.B., Panagopoulos, I., Thorsen, J. et al. (2012). T(12;13)(q14;q31) leading to HMGA2 upregulation in acute myeloid leukemia. *Br. J. Haematol.* 157: 762–774.

Pejovic, T., Heim, S., Mandahl, N. et al. (1989). Consistent occurrence of a 19p+ marker chromosome and loss of 11p material in ovarian seropapillary cystadenocarcinomas. *Genes Chromosomes Cancer* 1: 167–171.

Quade, B.J., Weremowicz, S., Neskey, D.M. et al. (2003). Fusion transcripts involving HMGA2 are not a common molecular mechanism in uterine leiomyomata with rearrangements in 12q15. *Cancer Res.* 63: 1351–1358.

Sait, S.N.J., Dal Cin, P., Sandberg, A.A. et al. (1989). Involvement of 6p in benign lipomas. A new cytogenetic entity? *Cancer Genet. Cytogenet.* 37: 281–283.

Sandberg, A.A. and Meloni-Ehrig, A.M. (2010). Cytogenetics and genetics of human cancer: methods and accomplishments. *Cancer Genet. Cytogenet.* 203: 102–126.

Seidal, T., Mark, J., Hagmar, B., and Angervall, L. (1982). Alveolar rhabdomyosarcoma: a cytogenetic and correlated cytological and histological study. *APMIS* 90: 345–354.

Slaughter, D.P., Southwick, H.W., and Smejkal, W. (1953). "Field cancerization" in oral stratified squamous epithelium. *Cancer* 6: 963–968.

Speleman, F., Dal Cin, P., de Potter, K. et al. (1989). Cytogenetic investigation of a case of congenital fibrosarcoma. *Cancer Genet. Cytogenet.* 39: 21–24.

Stenman, G. and Mark, J. (1983). Specificity of the involvement of chromosomes 8 and 12 in human mixed salivary-gland tumours. *J. Oral Pathol.* 12: 446–457.

Teixeira, M.R., Pandis, N., and Heim, S. (2002). Cytogenetic clues to breast carcinogenesis. *Genes Chromosomes Cancer* 33: 1–16.

Turc-Carel, C., Philip, I., Berger, M.-P. et al. (1983). Chromosomal translocations in Ewing's sarcoma. *N. Engl. J. Med.* 309: 497–498.

Turc-Carel, C., Dal Cin, P., Rao, U. et al. (1986). Cytogenetic studies of adipose tissue tumors. I. a benign lipoma with reciprocal translocation t(3;12)(q28;q14). *Cancer Genet. Cytogenet.* 23: 283–289.

Turc-Carel, C., Dal Cin, P., Boghosian, L. et al. (1988). Consistent breakpoints in region 14q22-q24 in uterine leiomyoma. *Cancer Genet. Cytogenet.* 32: 25–31.

Whang-Peng, J., Bunn, P.A., Kao-Shan, C.S. et al. (1982). A nonrandom chromosomal abnormality, del 3p(14-23), in human small lung cancer. *Cancer Genet. Cytogenet.* 6: 119–134.

CHAPTER 10

Gains, Losses, and Rearrangements of Genomic Material: Pathogenetic Considerations

Although a lot of evidence accumulated during the latter half of the preceding century showed that the central tenet of Boveri's somatic mutation theory of cancer was essentially correct (acquisition of the right kind of chromosome aberrations by susceptible, immature target cells leads to their neoplastic transformation), many questions remained unanswered, leaving our understanding of tumorigenesis very much incomplete. One aspect of this insufficiency regards what constitutes "susceptibility"; most researchers believe the susceptible target to be a stem cell of some kind, more or less committed in its preprogrammed pattern of differentiation, an opinion we share. This has not been unequivocally corroborated, however, at least not by data of a cytogenetic nature, so our stance in the matter is of little scientific value.

The second proviso in our rephrasing of Boveri's somatic mutation *credo* has to do with the "right kind of aberrations" that must occur for neoplastic transformation to ensue. Obviously, aberrations are of many different kinds,

Abnormal Chromosomes: The Past, Present, and Future of Cancer Cytogenetics.
Sverre Heim and Felix Mitelman.

with similarities as well as differences among them. It seems obvious that the *precise site* within the chromosome complement, or genome in more modern terminology, that is hit would be of importance, but it is at least likely that different rearrangements may lead to different outcomes even if they affect the same chromosome, chromosome arm or gene locus.

Acquired chromosome abnormalities can be subdivided in many different ways, limited only by the researcher's or taxonomist's imagination, but for pathogenetic considerations, a cytogenetic trichotomy comes across as logical: The aberrations may involve gain(s) of genomic material, loss(es) of material, or relocation of chromatin blocks without net gain or loss. The pathogenetic consequences envisaged in these three principal scenarios (Figure 10.1) differ, although some overlap exists that we shall revisit later.

The most thoroughly investigated situation involves balanced rearrangements, i.e., the movement of smaller or larger chromosome segments to a new position, leading to something happening at one of the two breakpoints, or possibly both, that unleashes neoplastic transformation. We have already written about this "something" in preceding chapters; it is the activation of

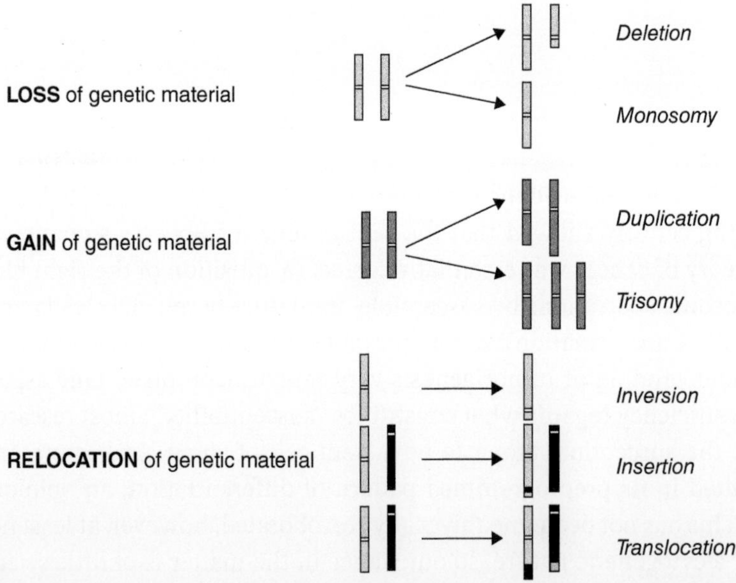

FIGURE 10.1 The chromosome aberrations of cancer may in principle exert their effect through gain or loss of genetic material or through structural or regulatory changes brought about by relocation of chromosomal segments (inversion, insertion or translocation).

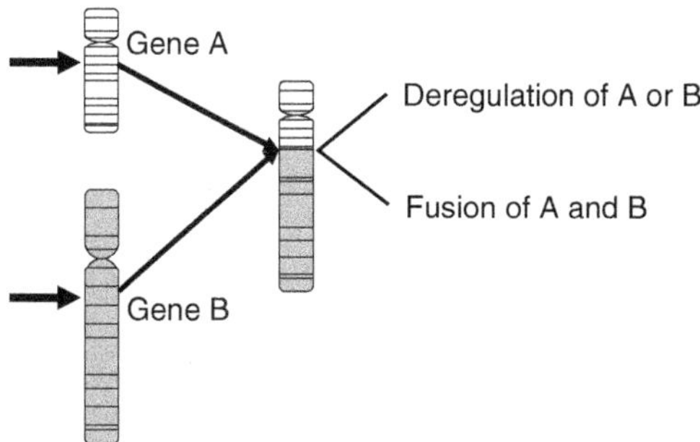

FIGURE 10.2 Consequences of balanced chromosome aberrations: Deregulation, usually overexpression, of a gene in one of the breakpoints or fusion of parts of two genes, one in each breakpoint.

a protooncogene into a fully fledged, positively acting (positive in the sense that we at least mostly have a gain-of-function alteration or mutation of the gene, not that it loses function) oncogene that produces an oncoprotein which, in its turn, causes all the protein-level biochemical changes transformation consists of. We have also pointed out that there are two principal types of oncogene activation through chromosomal means (Figure 10.2): Qualitative and quantitative. Qualitative oncogene activation – the t(9;22) of Philadelphia-positive leukemias giving rise to *BCR::ABL1* is the classic example – is typically brought about by the generation of a breakpoint fusion gene caused by a cancer-specific translocation, inversion or insertion. The altered and activated oncogene retains some of its original exons but also receives exons from the gene with which it fuses, leading to a qualitatively new primary gene structure which in its turn leads to an oncoprotein whose amino acid composition is novel and unique. Quantitative oncogene activation, on the other hand, typically exemplified by the *MYC* activation occurring as a consequence of translocations t(8;14), t(8;22), and t(2;8) in Burkitt lymphoma, refers to a situation in which the coding sequence of the oncogene remains intact, but its transcription is deregulated (usually enhanced) because the gene locus has come under the control of regulatory elements from the fusion partner gene.

Numerous examples exist of both qualitative and quantitative oncogene activation caused by chromosomal mechanisms (Heim and Mitelman 2015; Mertens et al. 2015; Rowley et al. 2015). We shall not probe any further into this subject here since it necessarily involves more molecular genetic than cytogenetic considerations, and the latter are our main topic in this book. Different techniques and resolution levels are optimally suited to answering different questions, not least when the mechanisms of tumorigenesis and leukemogenesis are concerned, and clearly cytogenetics has severe limitations when it comes to providing information about the very small, submicroscopic events. To answer molecular-level queries, molecular techniques are necessary.

We would, however, like to highlight two halfway cytogenetic exceptions from the standard rule of thumb that "balanced cancer-specific chromosome aberrations correspond pathogenetically to activation of oncogenes" we have just drawn up; biology is rich in possibilities and so is tumorigenesis, and it is not uncommon that many paths may lead to the same phenotypic end-point. Firstly and most importantly, a chromosomal rearrangement leading to gene fusion is not necessarily always balanced although it appears to be so by chromosome banding. Sometimes a deletion – be it visible by karyotyping or not – can cause fusion of two cancer-relevant genes, one located in each breakpoint (Mertens et al. 2015; Panagopoulos and Heim 2021). Perhaps the best-known example is the *SIL::TAL1* gene fusion that occurs through a submicroscopic deletion in 1p32 in a subset of T-lineage acute lymphoblastic leukemia. Secondly, and this exception is of a more noncytogenetic or rather subcytogenetic nature, it sometimes happens that a cancer-associated gene fusion does not lead to a genuine fusion gene, paradoxical though it may seem when written in this manner, but rather to a simple cutting off and shortening of the oncogene involved, a change which somehow has pathogenetic consequences. It may even be the case that balanced chromosome rearrangements result in gene truncations that lead to the inactivation or silencing of tumor suppressors (see more below), to isoforms with a dominant effect on the wild-type protein, or to haploinsufficiency (Mertens et al. 2015).

Thus, qualitative oncogene activation does not necessarily entail the acquisition of new structure and abilities. It nevertheless acts in a dominant fashion in the sense that the alteration affects only one of the two alleles in the transforming cell, and this is sufficient to wreak phenotypic havoc.

Questions of dominance or recessivity, time-honored genetic concepts since the days of Gregor Mendel, are even more relevant when it comes to the equally important question of how losses of genetic material – be they of entire chromosome copies, whole chromosome arms or deletions of smaller segments – act tumorigenically. It seems reasonable to enter the story of how losses work via a somewhat oblique angle, with the observation made by Henry Harris, Georg Klein, Eric Stanbridge, and others that tumorigenicity often manifests in a recessive manner in experimental systems (Stanbridge 1984). When cancer cells and nonmalignant cells are fused, the malignant phenotype of the former is typically suppressed, indicating that functional wild-type alleles at some chromosomal loci are capable of dominantly counteracting malignancy-inducing genetic changes. When the hybrid cells later gradually lose the normal chromosomes, as they are prone to do, the malignant phenotype reappears. A general argument that malignancy probably constitutes a recessive trait was also put forward around the same time by Ohno (1971) who suggested that subsequent genetic events could serve to unmask the recessive mutation through actual or functional loss of the remaining normal, wild-type allele.

Simultaneously, Alfred Knudson addressed the question of recessive cancer genes (we use the term "recessive genes" since almost everybody refers to genes in this manner, although strictly speaking it is not the genes themselves which are recessive or dominant but rather the phenotypic traits they control) on the basis of the age-specific incidences of cancer. For most malignancies, incidence rates increase with the power of age. Using retinoblastoma (RB) as a model, a cancer of the eye that typically affects children and may occur both sporadically and in a hereditary form, in 1971 Knudson formulated the two-step hypothesis for tumor development, a way of thinking that later was to have immense general influence on theories as to how losses of genetic material contribute to tumorigenesis. Both steps involved inactivation of antioncogenes or tumor suppressor genes, that is, genes capable of preventing a neoplastic phenotype from becoming manifest (Knudson 1983). The first step, theorized Knudson, corresponded to antioncogene inactivation either in the germline (in which case all somatic cells would carry the inactivated allele and, hence, be primed for possible later transformation) or in a somatic cell. The second step involved inactivation of the remaining wild-type gene so that both alleles now were dysfunctional in the cell undergoing transformation. Only after both antioncogene alleles are rendered inactive (homozygous inactivation) does the neoplastic phenotype become manifest and the cells begin to proliferate outside normal control.

Knudson's reasoning was primarily underpinned by cytogenetic, and later molecular genetic, findings in RB patients and their tumors. These studies also provided evidence that the inactivation events in a fair proportion of cases involve large enough stretches of chromosome material to be cytogenetically visible. About one-third of RBs are bilateral. All these tumors are hereditary (autosomal dominant transmission with 90% penetrance). Even in the unilateral group, 10% of cases are hereditary. Studies in the 1970s revealed that roughly 5% of RB patients have a constitutional deletion in the long arm of one copy of chromosome 13. The deletions varied in size but always involved chromosome band 13q14 (Yunis and Ramsay 1978); this must then be where the locus for the putative *RB* suppressor gene is situated, one concluded. Indeed, the gene in question did not remain putative for long before it was cloned (Friend et al. 1986), providing physical evidence that not only neoplasia-stimulating genes (oncogenes) but also neoplasia-suppressing genes were jointly controlling cell proliferation.

Studies of RB tumor cells provided partial corroboration of Knudson's two-step model requiring inactivation (by whatever means) of the relevant suppressor gene for neoplastic transformation to take place. Although acquired deletions of 13q are not commonly seen in RB cells (an isochromosome for 6p and rearrangements of chromosome 1, often leading to gain of 1q, are the most frequent aberrations), they do occur more often than chance would allow (Balaban et al. 1982). There are even some tumors in which cytogenetic evidence is found of homozygous 13q-deletions (Lemieux et al. 1989). Later molecular genetic investigations have corroborated this loss pattern surmised from cytogenetic data.

So it appears that inactivation of both *RB* alleles (the gene was later renamed *RB1*) is required for the neoplastic phenotype to become manifest, regardless of whether the disease is sporadic or hereditary. In the former situation, both inactivation events take place stochastically in the same somatic cell. Since this is highly unlikely, these patients do not have bilateral, let alone multiple, tumors. In the latter autosomal dominant scenario, however, all cells carry the first *RB1* inactivation whether this be cytogenetically visible or not. The likelihood of a second hit is then immensely larger than in individuals who have two competent alleles constitutionally, and so multiple or at least bilateral tumors are not rare.

Finally, a few words are due as to what the "inactivation events" may entail. We have already talked about submicroscopic mutations as well as deletions causing loss of the entire *RB1* locus, both of which may cause heterozygous loss of *RB1* function, but also nondisjunction, nondisjunction and reduplication, mitotic recombination and other forms of gene inactivation may ensure homozygous inactivation (Figure 10.3). At any rate, we are facing a pseudoparadoxical situation in which what comes across as a (sometimes) autosomal dominant cancer disease at the organism level is brought about at the level of individual cells by a recessive mechanism.

The double-hit model of Knudson, which was later shown to apply also to the osteosarcomas that often accompany eye tumors in RB patients, had a major impact on scientific thinking about tumor suppressor genes in general. Especially after *RB1* was first mapped and then cloned, so that everyone could see that this was not just a theoretical construct but a flesh-and-blood gene, enthusiasm soared. Would it be the case that all losses of genetic material in neoplastic cells – and remember, this was a very common

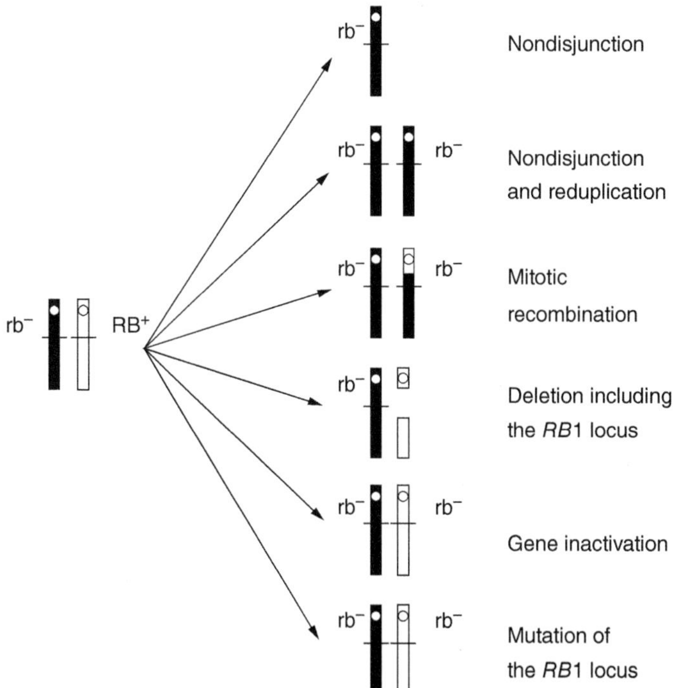

FIGURE 10.3 Mechanisms whereby hemizygosity or homozygosity for a defect in the retinoblastoma locus (*RB1*) may be achieved. The wild-type allele is designated RB^+, the defective allele rb^-. Inactivation of the wild-type allele in retinoblastoma can be achieved through nondisjunction with or without subsequent reduplication of the abnormal homologue, mitotic recombination, deletion of the wild-type suppressor locus, gene inactivation, or a second mutation. Source: Heim and Mitelman (1987).

situation – corresponded pathogenetically to loss of suppressor genes? Perhaps the correlation was even so specific that all antioncogene losses had to be homozygous for any phenotypic effect to become noticeable? Naturally, research interest focused primarily on the many highly characteristic losses seen in distinct malignancy types, such as deletions of 5q, 7q, and 20q in myeloid leukemias, deletion of 6q in lymphatic malignancies, of 3p in lung and kidney cancer, of 1p in neuroblastoma and so on. Did they all have in common inactivation, possibly homozygous inactivation, through loss of tumor suppressor genes there situated? If so, which were these genes?

Considerable efforts were put into this search by many highly competent research groups. This resulted in numerous scientific articles being published in even the most prestigious journals (we do not exactly favor the use of such adjectives in a scientific context, but admit to being massively outnumbered, perhaps even totally out of touch with the spirit of the times in this regard), reporting one partial success after another. However, it eventually turned out that it was extremely difficult to replicate in other tumor systems anything like the highly informative sequential discoveries related above for RBs. True, various tumor suppressor genes *were* detected in several cancers when the full power of molecular genetics was brought to bear; the most famous of them was *TP53* in 17p13 (Kastenhuber and Lowe 2017; Levine 2020) which may be inactivated by both large-scale deletions and much more subtle mechanisms. *TP53* is claimed to be the most important gene overall in carcinogenesis and its protein was in 1993 awarded the honor of being named "molecule of the year" by *Science* magazine, but the putative suppressor genes lost through deletions from 3p, 5q, 7q, etc. by and large remain elusive. Maybe they nevertheless exist as least common pathogenetic denominators in the malignancies carrying the said chromosomal aberrations – we should always be wary of interpreting absence of evidence as conclusive evidence of absence – but to this day they have proved beyond our abilities to identify.

It seems, therefore, that not all consistent losses of material by neoplastic cells are pathogenetically equivalent to loss of tumor suppressor genes, at least not according to the elegant two-hit scheme drawn up by Knudson and proven for RBs and some similar cancer types. The relatively rare generation of fusion genes in the deletion breakpoints alluded to above is probably not the only, let alone the most common, exception to the "rule" that loss of chromosome material is pathogenetically equivalent to loss of antioncogenes. Hemizygous loss of suppressor function may still be the explanation or other large-scale disturbances of gene regulation may work together with losses of important gene loci in situations where monosomies or deletions dominate the karyotypic picture, but the truth is that we still are not sure for most situations and tumors. That it has proved so difficult to deduce the functional consequences of many large-scale acquired genomic changes, such as losses of material, should definitely not lead us to believe that none exist. The cytogenetic aberrations are there because they confer on the cells having them some kind of evolutionary edge, otherwise we would not see them, and ours is the task to find out how this edge is achieved. As always, nothing makes sense in biology if not for evolution.

How do gains of genetic material contribute to tumorigenesis? The standard answer is that some kind of dosage effect must be operative with added (onco)gene copies in the gained material driving the cells toward longer life spans or a heightened mitotic rate. This all sounds good, albeit a bit vague. Furthermore, there is by no means any linear relationship between the number of alleles for a given gene locus, neither for increases to three copies in simple trisomies nor to many copies in cases of massive amplification, and the amount of protein produced by that particular gene (Chunduri and Storchová 2019; Ben-David and Amon 2020). Sometimes many copies yield more product, but not always. The fact is that we know little to nothing about how low-grade amplification, for example trisomy 8 in myeloid acute leukemia or trisomy 12 in chronic lymphocytic leukemia, leiomyoma, and benign ovarian tumors, alters the cells harboring the extra chromosome functionally, let alone why we see the type of disease specificity that evidently exists. It is presumed that all aberrations, including gains of whole chromosomes or parts of them, occur equally frequently in all neoplasms as they all represent stochastic disturbances of the normal state, but we do not actually know this. The skewed frequencies at which we observe them are presumably attributable to the magic of evolution that will always select the more fit cells making up a neoplastic clone, a theme we shall return to in the next chapter.

In summary, then, much of the relationship between acquired chromosome aberrations and the tumorigenic mechanisms they help unfold still remains largely incomprehensible to us although we do have some hard knowledge as well as some very tempting hypotheses as to how these examples of structure–function interaction may be explained. For many structural rearrangements, especially the balanced ones, qualitative or quantitative activation of oncogenes is the key, but in other situations we even today do not have a clue as to what is going on. Likewise, even extensive sequencing attempts often fail to detect fusion genes in the breakpoints of new translocations that we detect as part of our diagnostic cytogenetic work and then process for research purposes, and yet their very existence must mean that they somehow confer on the cells an evolutionary advantage. Obviously, there is a lot we still do not know.

This is even more striking for chromosomal aberrations involving loss of material, not to mention their opposite, karyotypic changes that involve gains of whole chromosomes, chromosome arms or smaller, yet microscopically visible segments. Although for the deletions and monosomies the functional gene-level equivalent is sometimes inactivation (through loss of one allele for a tumor suppressor gene), it is becoming increasingly difficult to believe that this is always the case. And for the chromosome aberrations corresponding to gains of larger or smaller genomic regions, we really are at a loss when asked for pathogenetic mechanisms.

We do not know, as already repeatedly pointed out, but perhaps larger-scale regulatory consequences develop when such chromosome-level mutations occur during tumorigenesis, a theme we shall return to later in this book. At the moment, we cannot do much more than paraphrase Shakespeare in a perhaps unseemly manner: There are more things in the biology of neoplastic transformation than we ever dreamt of in our philosophy. The science of life is so rich that it sends shivers down our spine even if we are allowed no more than microscopic glimpses of the intricacies involved.

REFERENCES AND FURTHER READING

Balaban, G., Gilbert, F., Nichols, W. et al. (1982). Abnormalities of chromosome #13 in retinoblastomas from individuals with normal constitutional karyotypes. *Cancer Genet. Cytogenet.* 6: 213–221.

Ben-David, U. and Amon, A. (2020). Context is everything: aneuploidy in cancer. *Nat. Rev. Genet.* 21: 44–62.

Chunduri, N.K. and Storchová, Z. (2019). The diverse consequences of aneuploidy. *Nat. Cell Biol.* 21: 54–62.

Friend, S.H., Bernards, R., Rogelj, S. et al. (1986). A human DNA segment with properties of the gene that predisposes to retinoblastoma and osteosarcoma. *Nature* 323: 643–646.

Heim, S. and Mitelman, F. (1987). *Cancer Cytogenetics*. New York: Alan R. Liss, Inc.

Heim, S. and Mitelman, F. (2015). *Cancer Cytogenetics. Chromosomal and Molecular Genetic Aberrations of Tumor Cells*, 4e. Oxford: Wiley-Blackwell.

Kastenhuber, E.R. and Lowe, S.W. (2017). Putting p53 in context. *Cell* 170: 1062–1078.

Knudson, A.G. Jr. (1983). Hereditary cancers of man. *Cancer Invest.* 1: 187–193.

Lemieux, N., Milot, J., Barsoum-Homsy, M. et al. (1989). First cytogenetic evidence of homozygosity for the retinoblastoma deletion in chromosome 13. *Cancer Genet. Cytogenet.* 43: 73–78.

Levine, A.J. (2020). p53: 800 million years of evolution and 40 years of discovery. *Nat. Rev. Cancer* 20: 471–480.

Mertens, F., Johansson, B., Fioretos, T., and Mitelman, F. (2015). The emerging complexity of gene fusions in cancer. *Nat. Rev. Cancer* 15: 371–381.

Ohno, S. (1971). Genetic implications of karyological instability of malignant somatic cells. *Physiol. Rev.* 51: 496–526.

Panagopoulos, I. and Heim, S. (2021). Interstitial deletions generating fusion genes. *Cancer Genom. Proteom.* 18: 167–196.

Rowley, J.D., Le Beau, M.M., and Rabbitts, T.H. (eds.) (2015). *Chromosomal Translocations and Genome Rearrangements in Cancer*. New York: Springer.

Stanbridge, E.J. (1984). Genetic analysis of tumorigenicity in human cell hybrids. *Cancer Surv.* 3: 335–350.

Yunis, J.J. and Ramsay, N. (1978). Retinoblastoma and subband deletion of chromosome 13. *Am. J. Dis. Child.* 32: 161–163.

CHAPTER 11

Morphology Meets Chemistry: Integration of Molecular Genetics into the Cytogenetic Search for Cancer-Specific Chromosome Aberrations

As scores of cancer-specific chromosome abnormalities were being detected during the latter half of the past century, another simultaneous approach to the study of human genomes in health and disease met with at least equal success: The molecular genetic analysis of neoplastic conditions. The findings obtained using the two methodologies – one based on microscopic examination of chromosomes arrested in metaphase and properly stained, the other relying on different types of chemical analyses of DNA, RNA, and, to a lesser extent, their protein products – cross-fertilized and augmented each other to a remarkable extent.

The circumstance that molecular genetics deals with smaller things than can be seen cytogenetically is in one sense trivial, although of course extremely important. We have in previous chapters repeatedly mentioned how molecular-level analyses were crucial in detecting many cancer-relevant

Abnormal Chromosomes: The Past, Present, and Future of Cancer Cytogenetics.
Sverre Heim and Felix Mitelman.

genes, including oncogenes located in the breakpoints of various leukemia-, lymphoma-, and sarcoma-specific balanced chromosomal rearrangements, be they translocations, inversions or insertions. Sometimes the oncogenes are activated qualitatively by fusion with a partner gene, in which case their primary structure becomes altered, eventually giving rise to a correspondingly altered oncoprotein, or the chromosomal rearrangement results in deregulation of the gene, leading to too little or (more typically) too much of an otherwise normal oncoprotein. Since this is biology, however, not logic, on yet other occasions simultaneous deregulation and qualitative, primary structure change take place.

At any rate, the detection of specific chromosome aberrations helped pinpoint the genomic sites of cancer-relevant genes whereupon molecular genetic studies led to an understanding of how they are involved in neoplasia. In addition, detailed examinations of cancer-specific losses of genomic material sometimes led to the detection of antioncogenes or tumor suppression genes. Though this was often the case, we also saw in the preceding chapter how, in many situations, equally diligent searches failed to reveal any consistent tumor suppressor loss corresponding to monosomies or tumor-specific deletions.

During the 1980s, the losses/gains of genomic material as well as the acquired balanced rearrangements of neoplastic cells became the objects of increasingly detailed molecular genetic studies. At first, this was mostly done in parallel with corresponding chromosome banding investigations of the same diseases, but soon many researchers came to rely almost exclusively on molecular-level techniques. There were some obvious advantages to this shift from a morphological to more chemistry-oriented investigative approaches, but something was also lost in the process. Of this – perhaps surprisingly, but then again maybe not – many of the new generation of scientists entering the field were not always cognizant, judging from how things played out.

In order to detect genomic losses, be they brought about by nondisjunction errors, imbalanced translocations, deletions or whatever, loss of heterozygosity (LOH) studies became quite popular. The idea was excellent and so was the technique: If an individual is constitutionally heterozygous for a given gene, i.e., the allele inherited from the father is different from that inherited from the mother, then both variants are present in his or her cells. If one of the two loci is lost in a neoplastic clone, however, perhaps because it was deleted, then one detects only the remaining allele. Such LOH could in earlier times be detected only at the protein level and then only in special situations, but now it became possible to do so directly, by examination of restriction fragment length polymorphisms (RFLP) present in the DNA itself. Numerous hitherto unknown investigative possibilities thus opened up, and researchers acted accordingly by subjecting lots of tumors and tumor types to LOH studies. Some new insights were certainly obtained, which is wonderful, but on other occasions the results merely confirmed already existing knowledge. There is nothing wrong in that, if only the fact had been appropriately recognized and the correct sequential order of discoveries acknowledged – independent confirmations of important findings by different techniques are of unquestionable value whenever they can be obtained – but unfortunately this was not always the case.

The gain/loss situation in meningiomas and similar mostly benign tumors is a case in point. Since prebanding times, it had been known that these tumors often lacked a G-group chromosome, an aberration later shown to correspond to monosomy 22. Now many studies appeared showing LOH of chromosome 22 markers (and in the beginning these markers were not many, it should be added), something that was hailed as a major new discovery. When this happens without proper referencing to the existing cytogenetic knowledge that loss of an entire chromosome 22 is common in meningiomas, however, something that *of necessity* must lead to LOH for whatever genes show heterozygous qualities on that chromosome, then the situation becomes at best awkward.

The same story repeated itself also for studies of other well-known neoplasia-specific losses, leading many cancer cytogeneticists to feel that their work was not given due credit. The take-home message is worthy of being generalized: To reference relevant findings obtained by researchers using other techniques than one's own should be an obvious requirement in scientific publishing, but sadly this does not always happen. The principle was sinned against in the 1980s and the same still happens to this day.

Another odd thing that discussions based on findings obtained using the LOH technique, even in published articles, was that LOH was often referred to, including by those who should have known better, as a pathogenetic mechanism. And yet this is clearly not so; LOH is an indicator, a mere marker that loss of genetic material has occurred, and it is this latter *loss* which presumably contributes to the process of tumorigenesis. Heterozygosity as such is irrelevant in the context.

In addition to techniques building on the LOH principle, the development of more and more refined *in situ* hybridization methodologies at the juncture between molecular genetics and cytogenetics led to the introduction of several new ways of detecting both unbalanced and balanced genomic alterations in neoplastic cells, even at very high resolution levels. Especially after it became feasible to use nonradioactive substances, in particular fluorescent dyes, as signal molecules (fluorescent *in situ* hybridization or FISH), this new way of probing for genomic abnormalities became increasingly important both scientifically and in clinical settings. In the early 1990s, the introduction of chromosomal comparative genomic hybridization (CGH) (Kallioniemi et al. 1992; Pinkel et al. 1998), which is based on the comparison of normal and tumor DNA hybridization patterns to a reference set of human chromosomes (stronger colors when the tumor DNA contains more than the normal copy number whereas the opposite is the case if loss of alleles has occurred), enabled reliable assessment of genomic imbalances in tumors. The resolution level of this method was largely on a par with that of banding cytogenetics, but now it was no longer necessary to induce cell division to obtain informative results, which was a major advantage.

On the other hand, the picture arrived at represented an average of all the genomes present in a sample, from parenchymatous and stromal cells alike, so in this respect CGH hardly was a cellular or cytological method in the classic sense of the word. Likewise, although both major types of imbalances (gains as well as losses) were detected, balanced rearrangements could not be seen. These complementary qualities of chromosome banding analysis and CGH-based techniques meant that to a considerable extent they canceled out each other's shortcomings. In short, the use of both added to the usefulness of chromosomal studies in a major way, scientifically as well as diagnostically.

Various array technologies became widespread shortly afterwards and so CGH analyses took another step away from cytogenetics when array-based comparative hybridizations (aCGH) (Figure 11.1) gradually took over from "classic" chromosomal CGH (cCGH; one can only wonder at how swiftly techniques presently are being relegated by having attached to them adjectives such as "classic" or "conventional" as they are supplanted by more fashionable approaches, regardless of whether the latter are wholly new or, as in this case, modifications of existing methods). The principal advantages and disadvantages of aCGH were the same as for cCGH, but the former method lent itself more readily to automation and was also clearly more accurate, not least in the way much smaller genomic imbalances could now be scored (Pinkel and Albertson 2005). Hence, aCGH quickly took over.

Simultaneously, fluorescent hybridization techniques using painting probes (that map to whole chromosomes or large chromosomal segments), repeat sequence probes (that map to centromeres, telomeres, and polymorphic satellites), and single-copy probes (that became easy to produce after the Human Genome Project's successful completion around the beginning of the new millennium; these probes can in principle be made to hybridize to any genomic locus) became available. In multicolor karyotyping, simultaneous

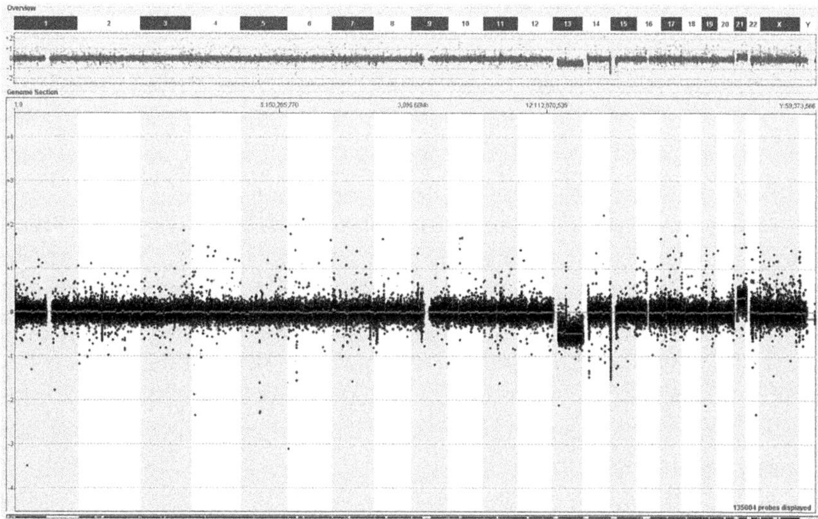

FIGURE 11.1 Array-based comparative genomic hybridization (aCGH) compares the tumor genome against a reference genome and identifies differences between the two in terms of gains and/or losses of genetic material. In this typical figure from our everyday laboratory practice (horizontal bars on top are numbered 1–22 and X and Y corresponding to positions along the X axis whereas gains and losses are shown as vertical deviations), imbalances are seen indicating loss of one copy of chromosome 13 with simultaneous gain of one copy of chromosome 21. Although balanced genomic changes cannot be seen by this technique, it constitutes an excellent complement to chromosome banding analysis.

hybridization of differently labeled painting probes allows chromosome identification by assigning each pair of homologous chromosomes a certain spectral signature (Schröck et al. 1996; Speicher et al. 1996). Alternatively, one can create color banding along the length of one or several chromosomes based on cross-species hybridization or microdissection of chromosome segments (Müller et al. 1998; Chudoba et al. 1999).

Not only did all this mean – and the list of technical improvements above is by no means exhaustive – a considerable expansion of the methodological repertoire cancer cytogeneticists can rely on, the newly introduced technological improvements really lay behind new discoveries as to which chromosomal changes characterize which neoplasms. Having said this, however, many of the new twists – the various ways of introducing color banding, for example – did not really contribute as much as one would perhaps expect; regular G-banding detected most that was to detect at the chromosomal level, and beyond that molecular methods had to take over anyway. And as far as clinical cancer cytogenetics was concerned, testing by CGH for genomic imbalances and by locus-specific probes for gene-level aberrations was in most situations sufficient. The latter probing asked specific questions of the neoplastic cells about whether gene rearrangements were present or not, including whether some particular gene loci were lost or amplified. For example, interphase analyses (iFISH) would typically be used to find out whether one of the common leukemia-specific translocations leading to gene fusion was present, whether *TP53* was lost through a 17p deletion, whether chromosomes 5 and/or 7 were deleted or lost in a myeloid leukemia, or whether *ERBB2* amplification was present in breast cancer cells or *MYCN* amplification had occurred in neuroblastoma, all of which might be important diagnostically, prognostically or therapeutically. Only less frequently would the chromosomes of metaphase plates be hybridized with the same probes, a more cumbersome procedure but one that might reveal more detail about the nature of the rearrangements present. Especially for us old-timers, however, the latter approach is particularly treasured, for we know from long experience that "chromosomes do not lie." Findings in even a single, well-spread metaphase plate showing impeccable hybridization may be more informative than knowing that 5%, 10% or 20% of interphase nuclei display a given signal pattern indicating this or that.

The above-mentioned percentages hint at a systematic difference between morphological and chemical studies and distinctions between what is biologically (and statistically) significant and what is not. When we perform iFISH analyses of presumably neoplastic cells, it can be prohibitively difficult to draw the line between signal patterns indicating the existence of small, genomically abnormal clones and methodological noise reflecting the inherent limitations of the technique. All laboratories therefore have statistical guidelines which, unfortunately, are not always able to remove the uncertainty in borderline cases. Particularly difficult in practice can be the decision on whether a clonal monosomy exists (do we see a high enough number of interphase cells showing only one copy of the marker in question?), but also the sporadic finding of a third marker constitutes an interpretation problem. Loss of something (the first example) could easily be due to technical issues, but also the apparent gain of one copy may be spurious; some cells may have duplicated genomes, and if two of four signals then overlap by chance, a signal pattern indicating trisomy would be the result.

The most reliable clinically important findings obtained by iFISH result from the use of appropriately constructed double fusion probes to detect cancer-specific rearrangements. Typically, the melting together of one green signal (gene A) and one red signal (gene B) in a fusion gives rise to two yellow signals corresponding to the two hybrid genes AB and BA, while the two nonfused signals (one red and one green) hybridize to the normal chromosome copies. Chance overlapping of signals can largely be ruled out if two yellow signals are observed in the same, and especially in several, interphase nuclei. Be all this as it may, the iFISH interpretation problems are more akin to the same challenge that clinical chemists face daily, but that rarely plague those of us who work in banding cytogenetics. Statistical guidelines are helpful, but our professional life is infinitely easier when we can make do without them.

In the choice of which techniques to choose for which purposes, the main consideration is often whether one wants screening ability but inherently less investigative resolution, or whether one goes for answers to only one or a few specific questions asked of the cells under examination. These are strategic considerations that will be discussed more fully in Chapter 16; they are of immense importance and considerable difficulty, and they are not going to miraculously fade away just by the introduction of new technological improvements. Suffice it to say at this juncture that a combination of investigative approaches is often the best choice; G-banding provides an overview of the genome that is unbiased but only informative above a certain level of organization. In addition, one may also want to probe by molecular means for the status at a few genomic points of interest (be they chromosomes, chromosome arms or gene loci) that experience has taught us are particularly crucial in the given clinical setting.

The said approach is now standard diagnostic practice in the diagnostic work-up when leukemias are suspected, and it functions just fine. It would not surprise us in the least if leukemia cytogenetics were to show the way once again as to what constitutes the optimal interplay between different investigative approaches in malignant diseases generally.

REFERENCES AND FURTHER READING

Chudoba, I., Plesch, A., Lörch, T. et al. (1999). High resolution multicolor-banding: a new technique for refined FISH analysis of human chromosomes. *Cytogenet. Cell Genet.* 84: 156–160.

Dal Cin, P., Qian, X., and Cibas, E.S. (2013). The marriage of cytology and cytogenetics. *Cancer Cytopathol.* 121: 279–290.

Kallioniemi, A., Kallioniemi, O.P., Sudar, D. et al. (1992). Comparative genomic hybridization for molecular cytogenetic analysis of solid tumors. *Science* 258: 818–821.

Müller, S., O'Brien, P.C., Ferguson-Smith, M.A., and Wienberg, J. (1998). Cross-species colour segmenting: a novel tool in human karyotype analysis. *Cytometry* 33: 445–452.

Pinkel, D. and Albertson, D.G. (2005). Array comparative genomic hybridization and its applications in cancer. *Nat. Genet.* 37 (Suppl): S11–S17.

Pinkel, D., Segraves, R., Sudar, D. et al. (1998). High resolution analysis of DNA copy number variation using comparative genomic hybridization to microarrays. *Nat. Genet.* 20: 207–211.

Ribeiro, I.P., Melo, J.B., and Carreira, I.M. (2019). Cytogenetics and cytogenomics evaluation in cancer. *Int. J. Mol. Sci.* 20: 4711.

Sandberg, A.A. and Meloni-Ehrig, A.M. (2010). Cytogenetics and genetics of human cancer: methods and accomplishments. *Cancer Genet. Cytogenet.* 203: 102–126.

Schröck, E., du Manoir, S., Veldman, T. et al. (1996). Multicolor spectral karyotyping of human chromosomes. *Science* 273: 494–497.

Speicher, M.R., Gwyn Ballard, S., and Ward, D.C. (1996). Karyotyping human chromosomes by combinatorial multi-fluor FISH. *Nat. Genet.* 12: 368–375.

Swansbury, J. (ed.) (2003). *Cancer Cytogenetics. Methods and Protocols*. Totowa: Humana Press.

CHAPTER 12

Unraveling the Clonal Evolution of Neoplastic Cell Populations

When abnormal chromosomes are found in clonal proportions in cells cultured from a neoplastic lesion, several questions come to mind: What does the finding signify, clinically (see next chapter) as well as in terms of information about the tumorigenic process? If more than one acquired aberration is seen, which is often the case, are all the observed changes equally important? Above all, how does the clonal composition of a neoplastic process evolve over time?

We have already made it clear that in cancer cytogenetics, we do not much concern ourselves with the many why-questions encountered when we try to understand neoplastic processes – how-questions are more than difficult enough for us – but we cannot completely avoid reflecting, however briefly, on why at least some of the observed aberrations come into being. As for primary chromosomal changes, those that often or at least sometimes occur as sole aberrations, the consensus holds that they mostly reflect stochastic events selected for by the fact that they transform neoplastically the cell(s) hit by them. The exception to this rule or theory is provided by some primary aberrations in secondary leukemias; these chromosomal abnormalities differ depending on which clastogen (a chemical able to induce structural chromosomal mutations is called clastogenic) caused the iatrogenic disease.

Abnormal Chromosomes: The Past, Present, and Future of Cancer Cytogenetics.
Sverre Heim and Felix Mitelman.

As far as secondary aberrations are concerned (neoplasia-associated chromosome aberrations are called secondary – and possibly even tertiary, quarternary, etc. to the extent that relevant sequential knowledge exists – when they develop in cells that have already been transformed by a preceding change), for them, too, the main hypothesis holds that they develop by chance. The fact that patterns of secondary changes typically differ depending on which primary aberration is present (the clonal evolution patterns in Philadelphia-positive leukemias are a case in point, albeit only one of many examples from both hematopoietic neoplasms and solid tumors) is presumably due to evolutionary selection, not induction of any kind. The subclone harboring the secondary abnormality providing the most "fitness" expands more readily in accordance with the Darwinian "survival of the fittest" principle, whereas any neutral additional aberrations remain rare (meaning that they are seen in only one or few cells) and deleterious ones quickly fade into relative oblivion. Thus arise the more or less complex changes we observe when neoplastic cell nuclei are studied with the purpose of detecting their chromosome constitution.

But in all likelihood there is more to the story than chance plus selection; at the very least, there have got to be several types of chance working on the observed clonal outcome after a given neoplasm has had time enough to modify and optimize its clonal composition. This should in theory apply regardless of whether a given tumor is benign or malignant, whether it was treated with toxic substances or not. There are several scenarios (Figure 12.1) that may be envisaged theoretically accounting for how neoplastic clones develop over time (Heim et al. 1988; Teixeira and Heim 2011), all of which have been observed in practice. A perusal of the possibilities may help us understand more fully which developmental tendencies are at work.

In the first scenario (Figure 12.1a), a single chromosomal abnormality exists from the time of diagnosis (actually from the time of transformation, only we could not observe the goings on at that time) and throughout the entire period of observation, i.e., until the patient dies, from his disease or with it, or for other reasons is no longer being monitored. It is important to emphasize that although this persisting simplicity may not be the most common situation, it is by no means exceptional. In leukemias, the same simple primary chromosome abnormality often remains the only visible genomic change through several relapses, and the same pathogenetically uneventful development – or rather lack of development – may also be observed in solid tumors, including sarcomas and even carcinomas. Examples of the latter are addition of extra material from the long arm of chromosome 1 as the sole chromosomal abnormality in endometrial carcinomas, often as an

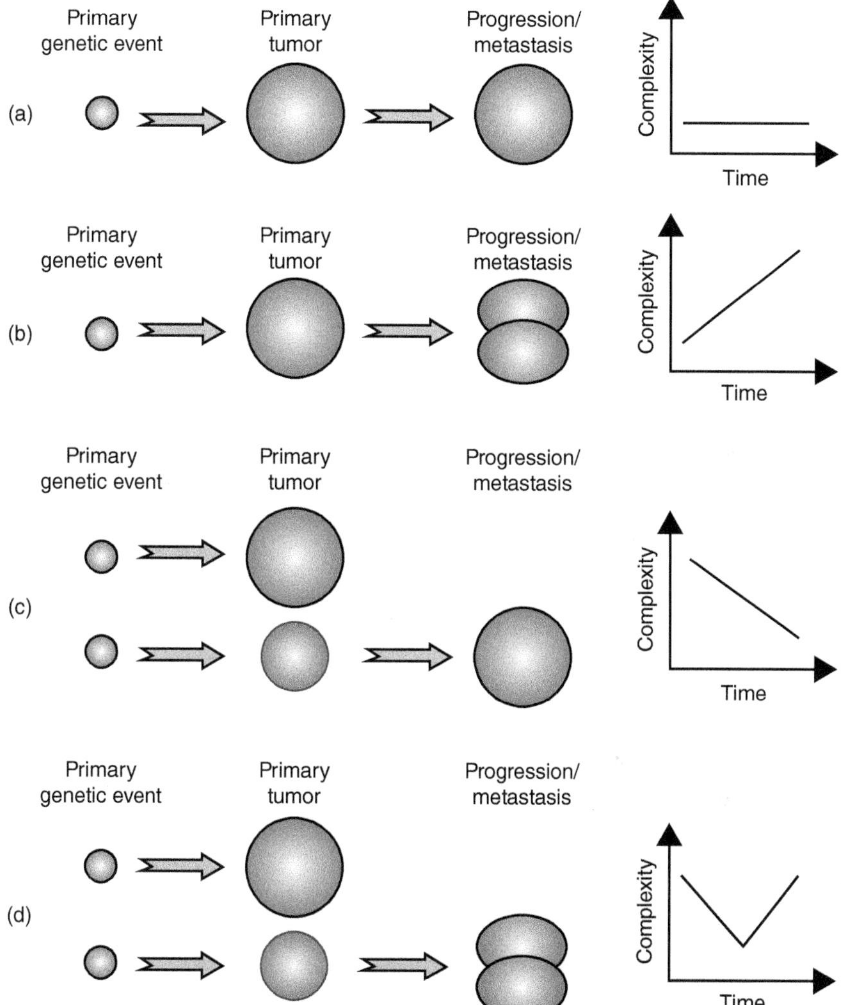

FIGURE 12.1 Genetic stability, divergence, and convergence during tumor progression. (a) Monoclonal tumorigenesis with genomic stability throughout tumor progression; no additional chromosome aberrations develop. (b) Monoclonal tumorigenesis displaying genetic divergence with time as the result of the acquisition and clonal selection for secondary cytogenetic changes (overlapping ellipses represent cytogenetically related clones). (c) Polyclonal tumorigenesis with an overall reduction in genomic complexity with time (clonal convergence) due to selection among cytogenetically unrelated clones during tumor progression and the development of metastases (the areas of nonoverlapping circles represent the relative size of cytogenetically unrelated clones). (d) Polyclonal tumorigenesis with initial clonal convergence followed by later clonal divergence due to the acquisition of additional cytogenetic changes by the clone that was selected during the development of metastases. Source: Teixeira and Heim (2011).

isochromosome for 1q, and t(X;18)(p11;q11) occurring as the sole change in synovial sarcoma or t(12;16)(q13;p11) likewise in myxoid liposarcoma.

The simplicity and stability over many years in the latter scenario compel us to take issue with some claims often heard about the genetic composition of malignant tumors. No, malignancies are not *always* characterized by genomic complexity – it is sometimes way too easily claimed that 4–6 transforming changes are necessary for a malignant phenotype to ensue – nor are they *always* genetically unstable leading relatively quickly to a wide array of secondary abnormalities becoming visible in already neoplastic clones. Some even highly malignant tumorous examples exist, as pointed out above, in which this simply is not so. Whether additional submicroscopic abnormalities accrue over the years is a moot point. Maybe they do, but maybe not, at least not always; we can with confidence talk only about what we see, what *is* rather than what is possible but unproven. Regardless, the fact remains that there exists a substantial group of tumors that, at the chromosomal level, exhibit no increased karyotypic instability leading to more and more chromosome abnormalities detectable with existing methods, and yet they are unquestionably malignant.

The most common scenario, however, is that we first observe one primary chromosome aberration that then in due course, often as the neoplasm acquires increasingly malignant phenotypic features, is accompanied by secondary changes (Figure 12.1b). This may lead to nothing more complex than many aberrations being present in the mitotic cells we examine, although sometimes we see extensive complexity corresponding to the existence of several subclones, related by the same primary abnormality or even having a few secondary changes in common, whereas they otherwise differ karyotypically. The precise nature of the complexities may provide a key to the order in which clonal evolution within the neoplastic cell population occurred, which additional changes came before or after the others. Whatever the sequence, the phenotypic transformation observed is, from a cancer cytogenetic point of view, normally the *result* rather than the *cause* of the increasing clonal complexity we detect while monitoring the disease, although we cannot rule out the possibility that the opposite may also at times be the case.

The interplay between environmental factors and a stochastically increasing tendency toward karyotypic complexity, even chromosomal chaos, can be observed in one particularly revealing situation: When cytostatic drugs are administered to a cancer patient. Such a pronounced change in the selection pressure facing the neoplastic cells may lead to a more primitive or original clone again becoming the one that is most "fit" among the many subclones "fighting" for evolutionary supremacy. There is nothing strange in such "fitness shifts" even if they should bring about an abrupt decrease in the clonal diversity (Figure 12.1c,d) of the neoplastic cell population; the situation is very similar to what happens when a complex infectious disease is treated with first one, then another antiinfectious drug resulting in different subsets of bacteria being eliminated. Perhaps the most meaningful analogy between chemotherapeutic treatment of cancer and antibiotic treatment of infections exists between antituberculosis treatment of sometimes encapsulated and inaccessible bacilli and dormant, nondividing cancer cells. Both are notoriously difficult to eradicate even when the most sophisticated combination regimens are relied upon. To phrase the situation in teleological terms, however logically suspect that might be: The infectious agents/cancer cells utilize evolutionary diversity to their advantage to survive and thrive.

Are there other situations where a decrease (Figure 12.1c,d) rather than an increase in clonal neoplastic complexity can occur? To answer this question properly, we need to take into consideration that such complexity may be of two principally different kinds. First, we may have many secondary aberrations with or without this leading to the existence of perhaps multiple subclones cytogenetically related by sharing at least the primary, but perhaps also other chromosomal changes. Second, another kind of complexity may exist in which we find many cytogenetically abnormal but seemingly unrelated cells in the same neoplastic sample. Is this a picture commonly seen in cancer cytogenetics?

Let us take the first thing first, what we may call secondary clonal complexity inasmuch as the process itself was initially monoclonal (it began with the neoplastic transformation of a single somatic cell) whereupon a whole range of secondary changes accrued leading to many subclones. We have already pointed out that a considerable narrowing down of this complexity may occur if cytotoxic treatment is introduced, but are there not also less artificial events in a typical cancer's developmental history that may cause something similar? Indeed there are, with at least two of them representing severe potential clonal bottlenecks. We shall take the natural life of a carcinoma as an example.

To begin with, imagine a group of newly transformed carcinoma cells sitting together *in situ*, i.e., in their normal, familiar locale somewhere in the body. Such not yet infiltrating malignant epithelial cancers live a shielded life; they sit on top of the basal lamina and thus are unexposed to the full strength of the body's immune system. When they eventually infiltrate through that barrier or membrane, there probably is a whole lot of chance involved, at least with regard to the cellular characteristics that were the most important while the budding tumor was still shielded from blood-borne immunological defensive capabilities. There is no reason to believe that the tumor cells' abilities to withstand the many attacks they now all of a sudden have to overcome in order not to be eradicated, should be miraculously identical to the paramount abilities that gave them a growth advantage when the cells were still above the basal lamina. Inevitably, therefore, and regardless of the purely stochastic elements that certainly also come into play when infiltration occurs, they now face a novel situation where totally new cellular traits are selected for. If genetic heterogeneity leading to phenotypic variation existed at the time of basal lamina penetration, new subgroups of cells will now hold an evolutionary edge and be selected for. What was excellent in the absence of immune system attacks may now be largely irrelevant or worse; the situation cannot be otherwise.

The second evolutionary Rubicon in a developing carcinoma's life comes when a small chunk of infiltrating tumor dislodges, by pure chance (probably) or because it is somehow particularly prone to set off on such a journey, into a small lymphatic or blood vessel and eventually ends up in a parenchymatous organ, perhaps a lung, liver or brain. If the little tumor embolus gets stuck there, infiltrates the thin vessel wall and begins to grow in this very foreign land, a metastasis has developed. Again, the situation involves a most severe change of environment, and we can safely assume that what were favorable character traits *in situ*, when infiltration occurred or during the subsequent intravasal journey, are not necessarily optimal now. If genetic diversity corresponding to phenotypic diversity still exists at this point, it follows that a new round of selection occurs, leaving in command only those clones or subclones that are well suited to grow not only in the face of immunological hostility but even in this hitherto unfamiliar soil.

The common denominator of the scenarios outlined above is that massive changes in selection pressure do occur during a cancer's natural life, to which must be added all the various damnations – from the tumor's point of view – that modern medicine may bring to bear. This may lead not only to increased clonal complexity but also the reverse, the latter especially at the moment of beginning infiltration and when metastases are set up. It was highly gratifying when evidence was obtained demonstrating that practice reflected these theoretical scenarios: For example, some breast cancer metastases were shown to be less karyotypically complex than their corresponding primary tumors (Pandis et al. 1994, 1998; Teixeira et al. 1996).

Such reduction of clonal complexity with time is normally inferred when you find a lower number of secondary aberrations upon examination of a later tumor sample. Could it also manifest as a reduced number of cytogenetic clones, even clones that seem to be developmentally independent inasmuch as they share no common chromosome aberration, the scenario you would expect if they were all descendants of a single transformed cell? Does this latter situation occur at all and, if so, what does it imply?

Indeed it does, as we learned to our initial surprise when, in the 1980s and 1990s, we began to undertake more systematic karyotypic examinations of epithelial tumors. Several of the common carcinomas, including malignancies of the pancreas (Gorunova et al. 1995), breast (Pandis et al. 1993; Teixeira et al. 2002), oral cavity (Jin et al. 1990), and skin (Mertens et al. 1991), often demonstrated cytogenetically unrelated clones (Figure 12.2), something that is rare in hematopoietic malignancies (Heim and Mitelman 1989) and connective tissue tumors (Gorunova et al. 2001). Does not this discovery of a mosaic of cytogenetically independent clones making up the neoplastic parenchyma fly in the face of Boveri's theory, now well established as an indisputable principle, of monoclonal tumorigenesis?

It is worth emphasizing that although cancer cytogenetics, in the opinion of most workers in the field, primarily is a tool that provides us with an opportunity to obtain both clinically useful information and scientifically valid insights into the pathogenesis of neoplastic processes, its ability to assess the clonal composition of karyotypically diverse cell populations is also considerable. This is because acquired chromosome aberrations serve as reliable clonality markers that are faithfully passed down to all daughter cells. To boot, the aberrations are in principle infinitely variable and hence extremely informative about clonal relationships (Teixeira and Heim 2011). Far more than chemistry-based techniques, especially those based on female X chromosome inactivation used so extensively during the latter half of the last century to study the clonal composition of neoplasias, the examination of chromosomal markers has proved singularly useful in our given context.

Can we be certain that the cytogenetically unrelated clones observed in many carcinomas are also biologically unrelated, in which case they would not be sharing some microscopically invisible, unifying genetic feature inherited from the first transformed mother cell? No, we cannot, for such a conclusion lies beyond the reach of the techniques hitherto available for the study of such questions; it is still possible that this putative invisible something does exist. But if that turns out to be the case, the situation would be one in which we are witnessing a new type of secondary karyotypic diversity totally beyond anything we are accustomed to from leukemias, lymphomas, and connective tissue tumors. Presumably, in this scenario, the primary, invisible transforming event as a side-effect induces in the cells carrying it pronounced genomic

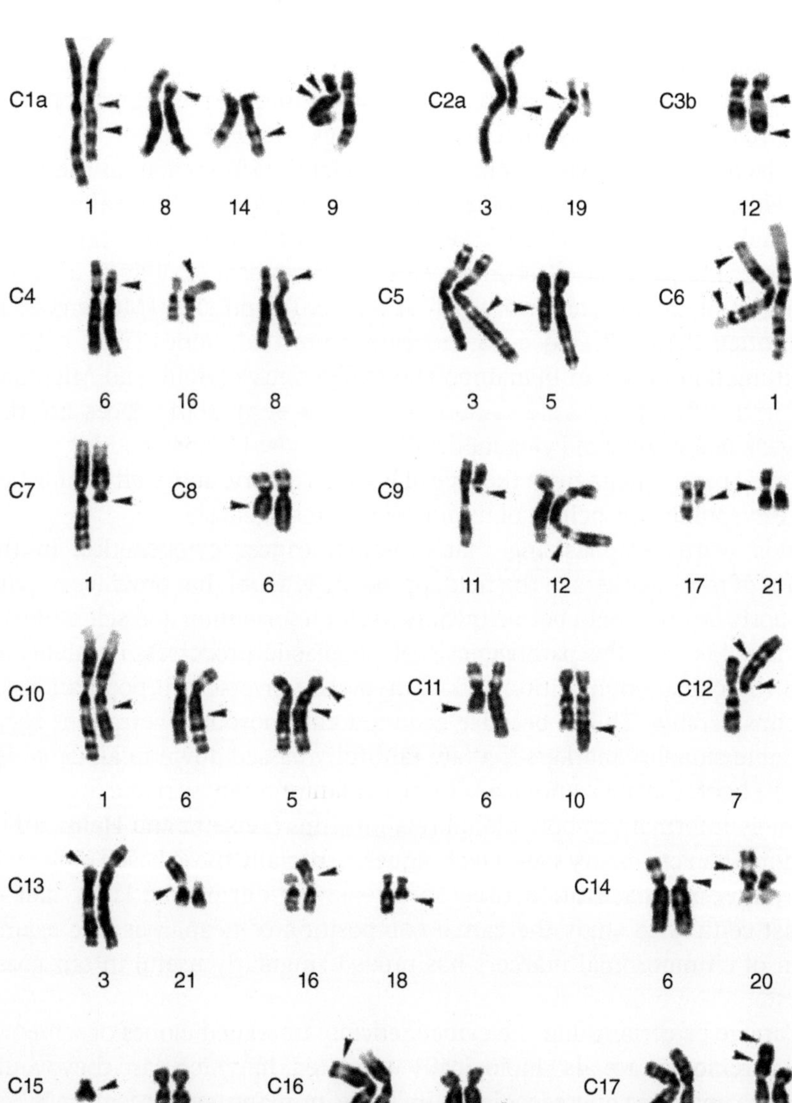

FIGURE 12.2 Partial karyotypes showing 17, designated C1–C17, out of a total of 54 unrelated clones found in a pancreatic adenocarcinoma. Arabic numerals (1–22) indicate chromosomes. Breakpoints in structural rearrangements are indicated by arrows. The presence of this many unrelated clones in the same tumor is exceptional – in fact, a world record – but the principle of cytogenetic polyclonality is well established in carcinomas of many organ systems. Source: Gorunova et al. (1995).

instability of a kind detectable by banding cytogenetics, which in its turn leads to the many disparate clones.

Such a sequence of events is undeniably possible, but does not strike us as particularly likely. In our mental universe, the jury is therefore still out discussing whether a major clonal difference in tumor development exists between some epithelial tumors and the rest of epithelial, connective tissue, and hematopoietic neoplasias, although we readily admit to holding a minority view in this regard. Beyond any doubt, the issue at some future point in time is going to be settled. We shall be comfortable with whatever the verdict will be as long as it is based on solid facts backed up by balanced, unprejudiced reasoning. For the phenomenon of extensive clonal heterogeneity does exist in many epithelial neoplasms, that much is certain. As always, chromosomes can be trusted.

Could it be that epithelial tumors more often are the result of carcinogenic exposures that are simultaneously strongly clastogenic, and that this clastogenicity brings about the many disparate clones? The thought is not unreasonable, but existing circumstantial evidence does not support it. We do find polyclonality sometimes in colorectal tumors, which seems to fit if we assume that carcinogens passed through the bowel system play a role in their genesis, but organs such as the breast and pancreas hardly qualify as particularly carcinogen exposed. Furthermore, tumors of the lung and urinary bladder – neoplasms whose development undoubtedly is stimulated by exposure to hazardous substances – do not often exhibit the above-mentioned polyclonality. Indeed, cytogenetic studies of anatomically distinct bladder tumors, even nonmalignant polyps occupying different locales within the same bladder, have shown them to be karyotypically related, indicating that they are but separate manifestations of the same neoplastic process (Fadl-Elmula et al. 1999). After our experience with other and more commonly polyclonal epithelial tumors, this came as yet another surprise for us, but that is practical research for you! Sometimes the findings do not fit the picture of reality you started out with, but scientific honesty always demands that the empirical facts be respected.

Finally, is there any theoretical reason why polyclonal tumorigenesis – whether the polyclonality be primary or secondary – should be more common in epithelia than in cells of other tissues? In trying to answer this question, we first assume that it is evolutionarily advantageous for the neoplastic process to consist of the many clones for which we see cytogenetic evidence; Darwinian thinking reigns supreme in all considerations about the clonal composition of neoplastic cell populations. This usefulness could then involve some kind of paracrine mechanism, that one cell secretes or otherwise contributes something that is of use for its neighbors and vice versa, something that helps them divide or expand more readily. We can further assume that such paracrine contributions to the expansion of (poly)clonal processes occur more readily when there is little distance among neighboring transformed cells, which is the very essence of how epithelia are organized compared with the much more scarcely populated bone marrow or connective tissues.

Maybe this line of thought contains something of relevance, maybe not. Many reasonable arguments turn out to be completely off topic in biology as well as other fields, both within the life sciences and otherwise. What remains certain, no matter what, is that relevant experimental facts must be taken into consideration when synthetic theories are construed as to what goes on. The cytogenetic polyclonality of epithelial tumors is one such observation that has been repeatedly corroborated; it certainly is no laboratory artifact, and it equally certainly does not fit with current conceptions of how neoplasms develop. We would dearly like to learn what lies hidden behind the puzzling facts while we still can.

REFERENCES AND FURTHER READING

Fadl-Elmula, I., Gorunova, L., Mandahl, N. et al. (1999). Cytogenetic monoclonality in multifocal uroepithelial carcinomas: evidence of intraluminal tumour seeding. *Br. J. Cancer* 81: 6–12.

Gisselsson, D. (2011). Intratumor diversity and clonal evolution in cancer – a skeptical standpoint. *Adv. Cancer Res.* 112: 1–9.

Gorunova, L., Johansson, B., Dawiskiba, S. et al. (1995). Massive cytogenetic heterogeneity in a pancreatic carcinoma: fifty-four karyotypically unrelated clones. *Genes Chromosomes Cancer* 4: 259–266.

Gorunova, L., Dawiskiba, S., Andrén-Sandberg, A. et al. (2001). Extensive cytogenetic heterogeneity in a benign retroperitoneal schwannoma. *Cancer Genet. Cytogenet.* 127: 148–154.

Greaves, M. and Maley, C.C. (2012). Clonal evolution in cancer. *Nature* 481: 306–313.

Heim, S. and Mitelman, F. (1989). Cytogenetically unrelated clones in hematological neoplasms. *Leukemia* 3: 6–8.

Heim, S., Mandahl, N., and Mitelman, F. (1988). Genetic convergence and divergence in tumor progression. *Cancer Res.* 48: 5911–5916.

Jin, Y.S., Heim, S., Mandahl, N. et al. (1990). Unrelated clonal chromosomal aberrations in carcinomas of the oral cavity. *Genes Chromosomes Cancer* 1: 209–215.

Kakiuchi, N. and Ogawa, S. (2021). Clonal expansion in non-cancer tissues. *Nat. Rev. Cancer* 21: 239–256.

Mertens, F., Heim, S., Mandahl, N. et al. (1991). Cytogenetic analysis of 33 basal cell carcinomas. *Cancer Res.* 51: 954–957.

Pandis, N., Heim, S., Bardi, G. et al. (1993). Chromosome analysis of 20 breast carcinomas: cytogenetic multiclonality and karyotypic-pathologic correlations. *Genes Chromosomes Cancer* 6: 51–57.

Pandis, N., Bardi, G., Jin, Y. et al. (1994). Unbalanced 1;16-translocation as the sole karyotypic abnormality in a breast carcinoma and its lymph node metastasis. *Cancer Genet. Cytogenet.* 75: 158–159.

Pandis, N., Teixeira, M.R., Adeyinka, A. et al. (1998). Cytogenetic comparison of primary tumors and lymph node metastases in breast cancer patients. *Genes Chromosomes Cancer* 22: 122–129.

Teixeira, M.R. and Heim, S. (2011). Cytogenetic analysis of tumor clonality. *Adv. Cancer Res.* 112: 127–149.

Teixeira, M.R., Pandis, N., Bardi, G. et al. (1996). Karyotypic comparisons of multiple tumorous and macroscopically normal surrounding tissue samples from patients with breast cancer. *Cancer Res.* 56: 855–859.

Teixeira, M.R., Pandis, N., and Heim, S. (2002). Cytogenetic clues to breast carcinogenesis. *Genes Chromosomes Cancer* 33: 1–16.

Turajlic, S., Sottoriva, A., Graham, T., and Swanton, C. (2019). Resolving genetic heterogeneity in cancer. *Nat. Rev. Genet.* 20: 404–416.

CHAPTER 13

Clinical Usefulness

One of the blessings of having a medical education is that one can make good use of it in many ways. For most medical students, the wish to help people certainly ranks high among the motivations that guide them through sometimes tough years. Another motivator of considerable importance – only for a minority, admittedly, but nevertheless – is scientific curiosity; how are the enormously diverse events of life, in particular human life in health as well as illness, developmentally strung together, even causally interconnected? How does one thing lead to another when disease strikes and which are the mechanisms behind pathological processes? For many medical doctors, us included, the possibility of being at the same time both clinically useful, at least to some extent, *and* involved in medical research stands as an ideal for a good professional life. Our work in cancer cytogenetics has helped us reach that goal.

Cancer cytogenetics has within clinical medicine first and foremost offered a new mode of diagnosing neoplastic diseases and subdividing them into meaningful entities. Secondarily, chromosome analyses have provided data that may be prognostically important, and which hence can help select optimal treatment for different groups of patients. Finally, with increased understanding of disease processes and knowledge about cancer-specific chromosome abnormalities and their molecular consequences, new, more effective and specific medications may be invented or detected that counteract the very genetic changes which brought about neoplastic transformation in the first place. The latter scenario long seemed nothing more than a dream or utopic vision, but its feasibility has lately been demonstrated for a few select groups of diseases and patients. The genetic approach to

Abnormal Chromosomes: The Past, Present, and Future of Cancer Cytogenetics.
Sverre Heim and Felix Mitelman.

oncological challenges is undoubtedly beginning to offer ever more personalized, effective medicine to the benefit of individual patients. The introduction of tyrosine kinase inhibitors to treat patients suffering from Ph-positive leukemias is the most poignant example of this principle, of medications interacting with the leukemogenic principle itself (see also below).

But we are getting ahead of ourselves; let us take first things first where clinical usefulness is concerned and begin with the diagnostic importance of chromosome analyses of potentially malignant processes. If normal cells never carry acquired clonal chromosome aberrations – the basic premise we work from – then the finding of at least one such change in clonal proportions should mean that the sampled disease process is of a neoplastic nature. However, cancer cytogenetics is part of biology and medicine, not mathematics, and so the said foundation is only almost always the case; a few notable exceptions are known to exist that demand our careful attention.

Older men tend to lose the Y chromosome from their cells, be it from the bone marrow or other tissues. Clonal loss of the Y therefore usually has no diagnostic information value, nor does it reveal anything about the prognosis of any given disease. But other numerical aberrations of sex chromosomes can be diagnostically challenging. Though they undoubtedly are sometimes acquired as part of a neoplastic process – this also goes for -Y – one always has to be careful not to misinterpret a cytogenetic change of questionable or no clinical significance or that is even inborn (presence of an extra copy of X or Y or loss of one X chromosome in women), possibly as part of constitutional mosaicism (particularly common in Turner syndrome). The possibility that an aberration might be constitutional, not acquired, is by no means restricted to sex chromosome changes, however. Whenever a rearrangement is found that is balanced, not typical of any known neoplasm, and/or present in all examined metaphase cells, one should see the warning lights blinking. More often than not, examination of a PHA-stimulated blood culture then reveals that the same change is present in T-lineage lymphocytes, too, meaning that it is a rare constitutional variant without clinical significance, at least not in an oncological context.

Other situations can be much more diagnostically intricate in the sense that the finding of some cytogenetically abnormal clones does not necessarily signify the presence of a neoplastic process. Perhaps the most common example of this difficulty is when small clones with autosomal trisomies, especially but not exclusively trisomy 7, are detected in tissue lesions. Experience has taught cancer cytogeneticists that this trisomy does not necessarily mean that the examined cells are neoplastic although, truth be told, we do not know for certain what they are. Some have suggested that they may belong to the immune system, but the evidence for this is at best circumstantial. Whatever their origin, we know that cells with +7 are sometimes found in unquestionably nonneoplastic disease processes (Mertens et al. 1996; Jin et al. 1997), while at the same time they have fairly frequently been detected in, amongst many other tumors, colorectal adenomas in a manner that would normally leave no doubt that the trisomy is then present in tumor parenchyma cells (Bomme et al. 1996).

Although most low-grade autosomal trisomies represent a diagnostic problem in solid tumor cytogenetics, the phenomenon is not completely unheard of in hematology, either. Sex chromosome aberrations were mentioned above, but also the importance of trisomy 15, occurring alone or together with -Y, has been called into doubt (Smith et al. 1999; Goswami et al. 2015). The bottom line is the following: Although the finding of an acquired, clonal chromosome aberration in a representative tumor sample nearly always proves that the patient has a neoplastic disease, some "quality control" of the aberration in question is necessary. This particularly involves ruling out the above-mentioned pitfalls, but is also important in some other cases (previously irradiated tissue lesions come to mind as an especially clinically relevant and difficult situation) before a definite answer can be given back to the pathologist or clinician.

It is important to emphasize that whereas the presence of a clonal abnormality means that a neoplastic disease is there, admittedly with the just-mentioned warnings, the fact that an abnormal karyotype is *not* found by no means rules out that a patient suffers from leukemia, lymphoma, solid cancer or some more benign neoplastic disease. Regular cautions against logical fallacies as well as standard caveats concerning inherently incomplete sensitivities of diagnostic techniques generally and wholly also apply within cancer cytogenetics. To make the point more explicit or concrete: Clinicians who ask for bone marrow cytogenetic analysis in order to "exclude the possibility" that their patient has myelodysplasia, leukemia or the like are asking something that the technique cannot deliver. As we have stated on more than one occasion, absence of evidence is not evidence of absence. So even if we disregard the possibility that leukemias and cancers might not have genomic changes visible at the karyotypic level (this is true for at least one-third of all acute leukemias), things may always be complicated by the fact that we may be looking at the wrong cells, perhaps even elements belonging to the stromal component.

The above argument applies most strongly when discussing the informative value of G-banding and similar analyses, although it also holds for interphase fluorescence *in situ* hybridization (FISH). Only in a few specific diagnostic situations involving a one-to-one relationship between disease phenotype and genotype can one conclude that a normal karyotype "guarantees" that the disease is not there. If no t(9;22) and/or *BCR::ABL1* gene fusion is detected by karyotyping, FISH or use of some suitable molecular genetic technique, the patient does not have chronic myeloid leukemia, nor does a patient have acute promyelocytic leukemia if no *PML::RARA* and/or t(15;17) is detected. But otherwise the general conclusion must be that whereas positive karyotypic findings are extremely informative, a normal karyotype does not tell us much in individual cases.

As we have seen repeatedly in the preceding chapters, cytogenetics can not only reliably confirm that a neoplastic condition exists, but also in many instances can determine which neoplasm is present; this ability flows from the fact that many neoplastic diseases, benign as well as malignant, are characterized by specific acquired chromosome abnormalities. As luck would have it, some phenotypically very close tumor types happen to have distinct karyotypic features, making chromosome analysis of the essence whenever diagnostic difficulties exist. Examples of such situations include distinguishing between highly differentiated liposarcomas and lipomas, between leiomyomas, endometrial stromal tumors and leiomyosarcomas of the uterine body, and many other differential diagnostic situations. One particularly important – and phenotypically difficult – scenario is when a child has a round cell, blue cell, small cell tumor, i.e., a tumor of such undifferentiated cells that standard histopathological examinations often end up with a considerable margin of uncertainty when attempting to reach a diagnosis. In this situation, tumor karyotyping may provide decisive input as to the various differential diagnostic possibilities (often they include Ewing sarcoma, rhabdomyosarcoma, mesenchymal chondrosarcoma, small cell osteosarcoma, hemangiopericytoma, and neuroblastoma) whenever a specific chromosomal abnormality is detected.

It is worth emphasizing that diagnostic difficulties may also arise in cases where none were expected to begin with. Often the initial expectation was that the case at hand would be plain sailing from the diagnostic point of view, and so perhaps care was not taken to obtain a suitable sample, be it from the bone marrow, an enlarged lymph node or a solid tumor that was removed, for chromosome banding analysis of the neoplastic cells. The small, initial window of diagnostic opportunity may, as a consequence, already be closed when difficulties later became apparent, leaving the patient to live, or die, with a diagnosis that was at best suboptimal. The lesson to be learned is simple, at least in principle: Whenever a malignant diagnosis is a possibility, the karyotype of that neoplastic clone should be determined. One never knows when the rare and unexpected may interfere with routine diagnostic practices, and to have established a neoplasm's karyotype may in such situations prove to be crucial. Without a precise, meaningful diagnosis, the choice of treatment can never be rational.

In addition to the ability of cytogenetics to provide a diagnosis of neoplastic disorders based on tumor karyotype, the cytogenetic data can also tell clinicians and patients a lot about prognosis, information that may later be translated into therapeutic decisions of considerable importance. This aspect of cancer cytogenetics is especially pronounced for the leukemias, the lymphomas, and bone and soft tissue tumors (Swerdlow et al. 2017; WHO 2020), but there are no theoretical grounds why cytogenetic information should not be similarly useful in other branches of solid tumor oncology, although the integration of cytogenetics into the clinical handling of such cancer cases is currently less advanced. In fact, a first attempt to incorporate cytogenetic abnormalities into the WHO classifications of solid malignancies other than bone and soft tissue tumors, namely malignant gliomas, was recently introduced (Louis et al. 2016).

Even in phenotypically identical leukemias, for example acute myeloblastic leukemia with maturation (AML-M2) or acute myelomonocytic leukemia (AML-M4), the two most common morphological subgroups of AML, it has been shown that several cytogenetic, and hence pathogenetic, pathways can lead to the same hematological results. Furthermore, it has turned out that whereas some AML subgroups are associated with a good response to antileukemic therapy (remissions are relatively easy to obtain and are typically of long duration), the opposite may be the case for other cytogenetic subgroups. Oftentimes it seems that the very number of chromosome aberrations is the deciding factor (complex leukemia karyotypes are the harbingers of an unfavorable outcome), although not rarely the prognostic–cytogenetic correlation is more specific. The same principle applies to the preleukemic myelodysplastic syndromes: Some cytogenetic subsets of MDS have relatively little tendency to progress to AML (5q− as the sole change is a case in point), whereas other do so quickly and seemingly inevitably. For example, patients with inv(3)(q21q26) or t(6;9)(p23;q34) may seek medical help when they are still at a myelodysplastic stage, but in all likelihood their disease progresses to an aggressive AML rather quickly and that leukemia in most instances does not respond favorably to standard AML treatment. Similar examples are plentiful among myeloid malignances, underscoring the following clinically significant principle: The diagnostic karyotype is crucial in order to assess the prognostic situation of a leukemia patient, in terms of determining the innate aggressiveness of the disease and, no less importantly, to select the treatment that practice has shown is best for this particular genetic subset. For some patients, bone marrow transplantation may offer the only realistic hope, others benefit most from treatment with various combinations of cytostatics, and so on.

The prognostic importance of the diagnostic karyotype is no less in lymphatic than myeloid malignances, be they chronic or acute. The answer to the question of whether to treat or not to treat chronic, initially low-grade diseases, such as multiple myeloma and chronic lymphocytic leukemia, depends largely on which cytogenetic abnormalities have accrued (FISH analyses play a major role here), and in acute lymphoblastic leukemia (ALL) the karyotype is crucial for risk stratification and, hence, the choice of treatment. The possibility of an eventual cure depends heavily on it. This situation is particularly acute in children: Whereas the most common cytogenetic subsets of pediatric ALL (presence at diagnosis of massive hyperdiploidy or an *ETV6::RUNX1* gene fusion brought about by a cytogenetically invisible 12;21- translocation) correspond to standard-risk leukemia that can be treated by relatively mild cytostatic regimens, several other but rarer chromosomal abnormalities signify high-risk disease that necessitates more intensive and toxic treatment (Figure 13.1). There are several important considerations hidden within these therapeutic decisions that should be made explicit.

First, acute leukemia is inherently a lethal disease that, in the absence of adequate therapy, is always deadly. Favorable or unfavorable cytogenetic features therefore make sense only against the background of state-of-the-art treatment that must at least sometimes be curative. It follows from this that whenever significant changes are made in the way a given serious disease such as leukemia is treated, one has to assess anew the prognostic significance of all markers, including the cytogenetic profile of the leukemic cells. What yesterday counted as a prognostically unfavorable genetic feature may tomorrow be unimportant or even a sign of better-than-average probability of cure. It cannot be emphasized enough that in this prognostic role, the leukemic karyotype is only an empirical marker of something, namely how likely it is that the patient may be cured using a particular treatment protocol, and empirical data change all the time. Thus, existing knowledge about which marker is favorable and which is not is in a state of never-ending flux.

To maintain a proper perspective on things, it is important to keep in mind that to eradicate the malignant cells from any given organism is always possible by either physical (surgery, irradiation) or chemical (cytostatics) means. The clinical balancing act consists in not damaging the patient too badly while doing it; one should not throw out the baby with the bathwater. While this dilemma may seem trivial in the most extreme cases, it is anything

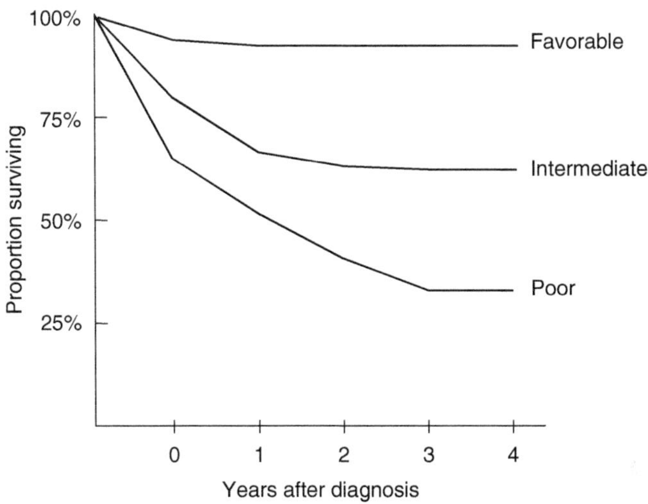

FIGURE 13.1 Early (1980s) stratification of childhood acute lymphoblastic leukemia cases into three distinct prognostic groups based on karyotypic aberration patterns detected at diagnosis. The "favorable" subset initially included cases characterized by massive hyperdiploidy but later, leukemias brought about by the submicroscopic t(12;21) leading to *ETV6::RUNX1* fusion were also shown to belong here. Several small genomic subsets corresponded to "unfavorable" leukemias, including cases characterized by t(9;22)(q34;q11) leading to *BCR::ABL1* fusion (for unknown reasons, this aberration is much more common in adult than childhood ALL). It is important to emphasize that all such prognostic classifications, then and now, are in a state of constant flux as they reflect outcome measured against the background of state-of-the-art therapy at the time of their making. If no effective treatment is available, ALLs of all genetic subgroups are rapidly lethal diseases, making the term "favorable" meaningless, even in its most comparative sense. On the other hand, a formerly "poor" prognosis leukemia subentity can later move into the "intermediate" or "favorable" category once better treatment regimens are introduced.

but when approaching the tipping point between too little effect on the cancer and too much damage to the patient. Yet again, the situation in childhood leukemia may serve as an example, perhaps even presaging the direction which other oncological developments are likely to take.

Whereas the prognostic focus of genetic analyses of leukemic cells in earlier years was primarily on identifying adverse prognostic subgroups whose patients would then receive tougher, and hence more effective, cytostatic treatment, now the emphasis is equally strong on identifying those leukemia patients who can be cured by relatively mild, although still quite taxing, protocols. Since medical doctors and their available supportive technology have become so extremely proficient in tiding cancer patients over the perils of acute cytostatic side-effects such as agranulocytosis and the life-threatening infections that often follow, few now die from therapeutic interventions during the early phases. However, after treatment of, for example, ALL in children became so efficient that most patients could be cured of their leukemia, more emphasis was gradually and rightly put on the late iatrogenic dangers that sometimes follow. These may involve not only impaired development of the child, but also an increased risk of developing secondary cancer or leukemia, and for both situations the more massive the cytostatic treatment has been, the higher is the risk. When only perhaps one-third of all patients survived their primary disease, these considerations rightly received less emphasis but now when more than 80% survive, the situation becomes very different. Whereas formerly the genetic markers looked at were primarily those that signified a need for more intensive treatment, nowadays it is no less important to identify those subsets of patients who are curable using treatment regimens that carry less severe long-term dangers.

Finally, the introduction of tyrosine kinase inhibitors in the treatment of patients with Philadelphia-positive leukemias – in particular patients with chronic myeloid leukemia (CML) – represented a watershed in our ability to target the primary acquired genetic abnormality of leukemic cells, in this case the *BCR::ABL1* fusion gene (Druker 2009; Braun et al. 2020). The drugs, initially imatinib but later numerous new variants, bind to the ATP-binding pocket of the *ABL1*, stabilizing the inactive, non-ATP-binding conformation of BCR::ABL1. This blocks the tyrosine kinase activity and inhibits both BCR::ABL1-mediated autophosphorylation and substrate phosphorylation, resulting in abrogated downstream signaling and reduced proliferation of *BCR::ABL1*-positive cells. Similar effects have been achieved in the much more rare leukemias harboring similar rearrangements affecting other members of the *ABL* family of kinase genes. The discovery of pathogenetic mechanisms that can be counteracted by means of tyrosine kinase inhibitor treatment has become a lively, albeit small, field within the genetic analysis of both hematopoietic malignancies and solid cancers.

The story of the Philadelphia chromosome, the 9;22-translocation, and the *BCR::ABL1* activated oncogene has thus entered a new and most telling phase, keeping this oldest of cancer-specific chromosome aberrations at the scientific forefront even as the continuous study of cancer genetics impacts the field of therapeutics (Figure 13.2). The CML situation provided not only proof of principle that a specific, highly effective drug can be produced that works directly against the very genomic abnormality that caused the malignancy, but showed that the principle worked in practice. Naturally,

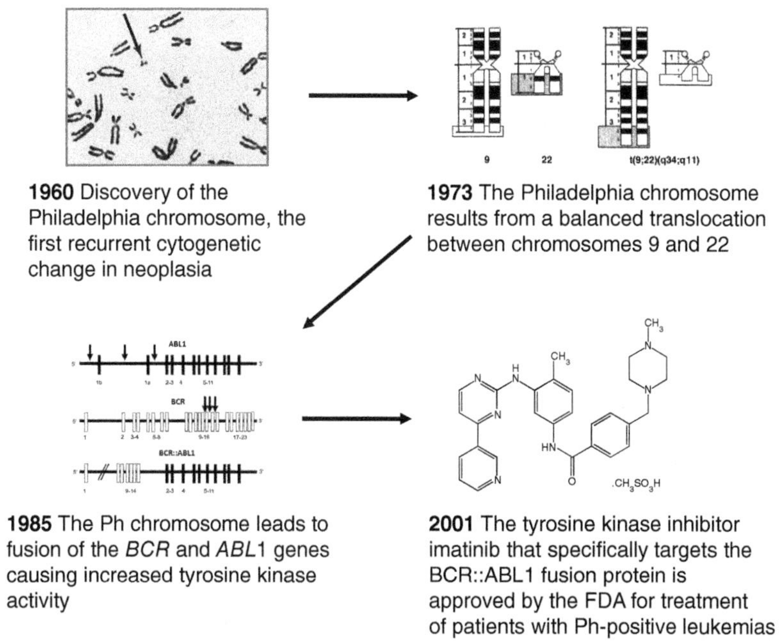
9
22
t(9;22)(q34;q11)
1960 Discovery of the Philadelphia chromosome, the first recurrent cytogenetic change in neoplasia
1973 The Philadelphia chromosome results from a balanced translocation between chromosomes 9 and 22
ABL1
BCR
BCR::ABL1
1985 The Ph chromosome leads to fusion of the BCR and ABL1 genes causing increased tyrosine kinase activity
2001 The tyrosine kinase inhibitor imatinib that specifically targets the BCR::ABL1 fusion protein is approved by the FDA for treatment of patients with Ph-positive leukemias

FIGURE 13.2 Historic milestones in the development of the first anticancer therapy based on cytogenetic knowledge: In 1960, the Philadelphia chromosome was detected as the first specific chromosome abnormality in a malignant disorder, chronic myeloid leukemia. In 1973, the Ph marker was shown by chromosome banding analysis to be the der(22)t(9;22) result of a balanced 9q34;22q11-translocation. In 1985, the translocation was shown to produce a *BCR::ABL1* fusion gene which, in its turn, gave rise to a corresponding fusion protein. Finally, imatinib treatment directed against the abnormal tyrosine kinase thus generated was approved in 2001 and revolutionized treatment as well as life expectancy of patients suffering from Ph-positive leukemias.

cancer cytogeneticists as well as oncologists are eager to follow this up with new and similar success stories, some of which are bound to occur in nonleukemic disorders. An example is the targeted treatment with crizotinib of a subtype of nonsmall cell lung cancer whose cells carry an *EML4::ALK* fusion brought about by an inversion in the short arm of chromosome 2 (Heigener and Reck 2018).

The chronologies of such discoveries are impossible to predict. It does seem very promising, however, that while it took 40 years after the discovery of the Philadelphia chromosome in CML to develop imatinib treatment, it took only a few years from the discovery of *EML4::ALK* fusion in lung cancer to the development of the tyrosine inhibitor crizotinib.

We have no doubt that new and similar discoveries and inventions will see the light of day in the years to come although it is not going to be easy. The cancer cytogeneticists' task in the greater picture of things is to find new cancer-specific genomic rearrangements, whereupon clever people of a more chemical bent stand ready to translate the new knowledge thus generated into therapeutic success.

REFERENCES AND FURTHER READING

Baliakas, P., Jeromin, S., Iskas, M. et al. (2019). Cytogenetic complexity in chronic lymphocytic leukemia: definitions, associations, and clinical impact. *Blood* 133: 1205–1216.

Bomme, L., Bardi, G., Pandis, N. et al. (1996). Chromosome abnormalities in colorectal adenomas: two cytogenetic subgroups characterized by deletion of 1p and numerical aberrations. *Hum. Pathol.* 27: 1192–1197.

Braun, T.P., Eide, C.A., and Druker, B.J. (2020). Responses and resistance to BCR-ABL1-targeted therapies. *Cancer Cell* 37: 530–542.

Chilton, L., Harrison, C.J., Ashworth, I. et al. (2017). Clinical relevance of failed and missing cytogenetic analysis in acute myeloid leukaemia. *Leukemia* 31: 1234–1237.

Döhner, H., Estey, E., Grimwade, D. et al. (2017). Diagnosis and management of AML in adults: 2017 ELN recommendations from an international expert panel. *Blood* 129: 424–447.

Druker, B.J. (2009). Perspectives on the development of imatinib and the future of cancer research. *Nat. Med.* 10: 1149–1152.

Goswami, R.S., Liang, C.S., Bueso-Ramos, C.E. et al. (2015). Isolated +15 in bone marrow: disease-associated or a benign finding? *Leuk. Res.* 39: 72–76.

Heigener, D.F. and Reck, M. (2018). Crizotinib. *Recent Results Cancer Res.* 211: 57–65.

Heim, S. and Mitelman, F. (eds.) (2015). *Cancer Cytogenetics. Chromosomal and Molecular Genetic Aberrations of Tumor Cells*, 4e. New York: Wiley-Blackwell.

Herold, T., Rothenberg-Thurley, M., Grunwald, V.V. et al. (2020). Validation and refinement of the revised 2017 European LeukemiaNet genetic risk stratification of acute myeloid leukemia. *Leukemia* 34: 3161–3172.

Jin, C., Jin, Y., Wennerberg, J. et al. (1997). Clonal chromosome aberrations accumulate with age in upper aerodigestive tract mucosa. *Mutat. Res.* 374: 63–72.

Johansson, B., Heim, S., Mandahl, N. et al. (1993). Trisomy 7 in nonneoplastic cells. *Genes Chromosomes Cancer* 6: 199–205.

Larson, D.P., Akkari, Y.M., van Dyke, D.L. et al. (2021). Conventional cytogenetic analysis of hematologic neoplasms: a 20-year review of proficiency test results from the College of American Pathologists/American College of Medical Genetics and Genomics Cytogenetics Committee. *Arch. Pathol. Lab. Med.* 145: 176–190.

Lazarevic, V.L. and Johansson, B. (2020). Why classical cytogenetics still matters in acute myeloid leukemia. *Expert. Rev. Hematol.* 13: 95–97.

Louis, D.N., Perry, A., Reifenberger, G. et al. (2016). The 2016 World Health Organization classification of tumors of the central nervous system: a summary. *Acta Neuropathol.* 131: 803–820.

Manier, S., Salem, K.Z., Park, J. et al. (2017). Genomic complexity of multiple myeloma and its clinical implications. *Nat. Rev. Clin. Oncol.* 14: 100–113.

Matsukawa, T. and Aplan, P.D. (2020). Clinical and molecular consequences of fusion genes in myeloid malignancies. *Stem Cells* 38: 1366–1374.

Mertens, F., Pålsson, E., Lindstrand, A. et al. (1996). Evidence of somatic mutations in osteoarthritis. *Hum. Genet.* 98: 651–656.

Moorman, A.V., Enshaei, A., Schwab, C. et al. (2014). A novel integrated cytogenetic and genomic classification refines risk stratification in pediatric acute lymphoblastic leukemia. *Blood* 124: 1434–1444.

Palumbo, A., Avet-Loiseau, H., Oliva, S. et al. (2015). Revised international staging system for multiple myeloma: a report from international myeloma working group. *J. Clin. Oncol.* 33: 2863–2869.

Perrot, A., Lauwers-Cances, V., Tournay, E. et al. (2019). Development and validation of a cytogenetic prognostic index predicting survival in multiple myeloma. *J. Clin. Oncol.* 37: 1657–1665.

Shah, V., Sherborne, A.L., Walker, B.A. et al. (2018). Prediction of outcome in newly diagnosed myeloma: a meta-analysis of the molecular profiles of 1905 trial patients. *Leukemia* 32: 102–110.

Smith, A., Watson, N., and Sharma, P. (1999). Frequency of trisomy 15 and loss of the Y chromosome in adult leukemia. *Cancer Genet. Cytogenet.* 114: 108–111.

Sonneveld, P., Avet-Loiseau, H., Lonial, S. et al. (2016). Treatment of multiple myeloma with high-risk cytogenetics: a consensus of the international myeloma working group. *Blood* 127: 2955–2962.

Swerdlow, S.H., Campo, E., Harris, N.L. et al. (eds.) (2017). *WHO Classification of Tumours of Haematopoietic and Lymphoid Tissues*, 4e. Lyon: IARC Press.

Wapner, J. (2014). *The Philadelphia Chromosome: A Genetic Mystery, a Lethal Cancer, and the Improbable Invention of a Lifesaving Treatment*. New York: The Experiment.

WHO (2020). *WHO Classification of Tumours. Soft Tissue and Bone Tumours*, 5e. Lyon: IARC Press.

Wolman, S.R. and Sell, S. (2011). *Human Cytogenetic Cancer Markers*. Totowa: Humana Press.

PART III

FUTURE

CHAPTER 14

Toward a Pathogenetic Classification of Cancer

In these final chapters of our story about the abnormal chromosomes that are so important in malignant as well as benign neoplasms, we shall attempt to share some insights into how cancer cytogenetics is likely to develop in the years to come. This is obviously a very tall order indeed, so before we begin in earnest, it seems prudent to reflect more generally on the legitimacy of such attempts at foreseeing the future. Two extreme positions can be discerned: Some people see guesswork of this type as inherently impossible and, hence, a total waste of time, while others maintain that knowledgeable people who are well versed in the subject matter ought to be able to predict what is going to happen with some precision.

Historicism is a nineteenth-century term that became popular following the postwar publication of Karl R. Popper's *The Poverty of Historicism* (Popper 1957). In it, the antitotalitarian Austrian-British philosopher debunked as fundamentally flawed the idea that we can fully comprehend the past as well as what is going to happen in the future through intimate knowledge of a set of laws that govern history. Such laws are nonexistent, Popper insisted, and we agree with him: They do not exist in politics nor do they exist in the empirical sciences, including medicine and cancer cytogenetics. Any statement about future developments has to keep this principal limitation in mind.

This does not mean that everything is equally likely, however, and there are at least two reasons why this is so. First, despite its shaky theoretical foundations, inductive knowledge does exist and is of value, as experience

Abnormal Chromosomes: The Past, Present, and Future of Cancer Cytogenetics.
Sverre Heim and Felix Mitelman.

tells us. Yes, we know that the above argument is painfully circular and hence well worthy of being branded the "scandal of philosophy," but nevertheless, we have all witnessed on many occasions that extrapolation from a set of data points into the future can give us a pretty good idea about what will be and what not. Medical research is a part of science, not philosophy or logic, and practice then always reigns supreme, not theoretical considerations. Second, an as profound as possible understanding of the field under scrutiny, including the relationships among its various elements, for example in the cancer cytogenetic context the hierarchical relations between genes and chromosomes, cellular genotypes and phenotypes and so on, can serve as a good indicator of what may lie ahead. It certainly will never guarantee the correctness of any detailed anticipations of future discoveries, of course not, but at the very least, we may have some inkling about the principal frameworks of tomorrow's science and medical practices. This approach is anyway better than being completely awestruck with every new discovery, for events of the past, present, and future are interconnected in this little part of research-driven medicine.

Why is it important by what names we call diseases, including the malignant ones (Brandal et al. 2010)? Because naming the various phenomena we encounter in life, especially the threatening ones, helps us in our striving to understand the outside world, even to obtain some kind of mastery over it. As for so much in philosophy and the natural sciences, it was Aristotle who in the *Organon* first thought and wrote systematically about what things should be called and how they should be classified into categories depending on their likenesses and differences. Especially important was the distinction he made between universals and particulars: Whereas individual, particular men can be dissimilar in many respects, all nevertheless share those universal qualities that make up the putative essence of "Man." Subsequent thinking has built on the foundation he laid. The phenomena of medicine, too, including tumors and other neoplastic processes, can be classified according to the Aristotelian scheme of things. When doing so, the grouping into entities, classes or categories should be based on their general (essential) as well as particular (primary and secondary) qualities.

To find out which are which, not least when different cancers and benign neoplastic processes are concerned, is of the utmost importance for our basic understanding of the disease processes, but also for prognostication and, in more and more situations as medicine advances, for the choice of therapy. It is mostly no longer sufficient to diagnose a condition merely as a cancer or benign tumor, for some cases of tumors or leukemias are sufficiently similar among themselves yet distinct from the remainder to merit grouping together

FIGURE 14.1 A famous illustration of how difficult it is to give an adequate description of any complex object depicts a group of wise but blindfolded philosophers while they are in the process of describing the defining features of that most majestic of animals, the elephant. Naturally, their description differs depending what part of the giant they lay their hands on: A leg is cylindrical, an ear is flat, and so on. Those who give names to complex diseases such as cancers face an even tougher task: Not only are they called upon to group disease instances according to macroscopic and microscopic phenotypic characteristics, they also increasingly have to consider genetic (at both the chromosomal and genic level) information in their classificatory schemes. Indeed, the latter information about the pathogenetic mechanisms at work is in recent years being seen as more and more important – essential even – in defining diagnostic entities and subgroups. This shift in emphasis is likely to continue. Source: Heim and Mitelman (2008).

in meaningful subcategories. How many groups should there be, and which criteria are decisive when exclusion or inclusion into a particular diagnosis (Figure 14.1) is decided?

Until a few decades ago, phenotype was the only thing to go by in these decisions. We have previously mentioned that brisk or slow-onset disease dichotomizes infections as well as leukemias (acute and chronic). As far as solid tumors are concerned, we have, besides the clinically crucial malignant/benign divide, tumor site, which microscopic tissue resemblance the new growth exhibits, the size and shape of tumor parenchyma cells, etc. that collectively determine the diagnosis. With the advent of cancer cytogenetics and molecular genetics, however, data on the neoplastic cells' acquired genomic changes are increasingly included in the equation. As repeatedly mentioned in preceding chapters, the tale of the Philadelphia chromosome developed into a more extensive and detailed story about how the *BCR*::*ABL1* fusion gene generates an oncogenic hybrid protein that can be counteracted by means of tyrosine kinase inhibitors. This paved the way for a new mode of thinking. Genetic classifications of leukemias and, by inference, cancers in general thus gained in importance (Swerdlow et al. 2017; WHO 2020). In other words, pathogenetic classifications gradually came across as more relevant than grouping neoplasms by secondary features such as cellular size or form, and the newfound therapeutic usefulness of genetic classifications (imatinib treatment of patients with Ph-positive leukemias) seemed to clinch the matter.

No matter which classification principle one adheres to, however, there will always be cases that do not quite fit in; this is almost a law of nature, it seems. Also, the increasingly relied upon (cyto)genetic classification of neoplastic processes contains numerous pitfalls or shortcomings exemplifying how difficult it may be to grasp both phenotype and genotype with one meaningful term. We outline below some of the principal challenges encountered.

It is not difficult to find situations in which one and the same cytogenetic aberration occurs nonrandomly in obviously different neoplastic diseases. Take some of the common trisomies as examples, say trisomy 8 or trisomy 12. The former is the most common chromosomal aberration in acute myeloid leukemia (AML), both occurring alone and in a secondary role, but obviously many other aberrations may be seen in leukemias of exactly the same phenotype. Furthermore, cancer cytogeneticists have for decades found trisomy 8 in a wide variety of non-AML/non-myelodysplastic syndrome (MDS) neoplasias, both of the bone marrow and solid tissues, so clearly +8 does not occur exclusively in hematopoietic neoplasms exhibiting myeloid differentiation.

Trisomy 12 is often found in three neoplastic conditions: Leiomyomas, benign and borderline ovarian tumors, and chronic lymphocytic leukemia. It is very hard to see any meaningful similarity between the three, at least between the chronic leukemia and solid tumors. Only genotypic classification fanatics, if any such there be, would insist on grouping together for clinical purposes all three neoplasias based on their frequent karyotypic similarity.

To this one might counter, for the sake of argument and upholding the extremist view of the putative fanatic envisaged above, that extra copies of an entire chromosome is a much too crude aberration to be given a decisive say in classification matters. But what if we increase the investigative resolution level from chromosomes to genes? Maybe then the same acquired aberrations turn out to correspond more reliably to distinctive phenotypic patterns, with fewer false negatives/positives messing up the picture? After all, we do not know how chromosome gains contribute to tumorigenesis, if at all, so maybe they themselves are effects rather than causes in neoplastic transformation?

Although there may be considerable principal validity to this argument, the one-to-many problems are not so "kind" as to just fade away when we move from larger to smaller organizational levels; it remains a fact that some oncogene-activating fusions are found in phenotypically very different neoplasms. The first example of this effect was the observation that t(9;22)(q34;q11) leading to *BCR::ABL1* was not exclusive to CML but also occurs in some acute, myeloid as well as lymphoblastic, leukemias. However, these are all hematopoietic neoplasias, one may argue, they do not develop from completely different cells and tissues, even from different germ layers. The same can be said about the multiple (>140 as of today) *KMT2A (MLL)* rearrangements, brought about by translocations targeting 11q23, that occur in AML and acute lymphoblastic leukemia (ALL) (Harper and Aplan 2008; Marschalek 2015). Some of them are more common in myeloid acute leukemia, others in lymphoblastic, whereas others show no distinctive differentiation preference. The prognostic impact associated with each *KMT2A* rearrangement varies to some extent, although for the less common subsets, the data are far from solid. To claim that all leukemias displaying *KMT2A* splitting are clinically more or less the same is clearly an oversimplification. The same also goes for other leukemia groups in which promiscuous oncogenes are involved, the many subsets showing *ETV6* rearrangements being the most prominent example.

Speaking of *ETV6* (12p13), we are here dealing with a gene that, when recombined with some translocation partners, appears to be the driving mutation behind several completely different (in the traditional, phenotypic sense) cancers. The most illustrative partner in this regard is the gene *NTRK3* in 15q25 which, through the translocation t(12;15)(p13;q25), may fuse with the 12p gene to bring about an *ETV6::NTRK3* hybrid. Though this malignancy-associated rearrangement is far from common, it has been seen in such thoroughly dissimilar conditions as acute myeloid and lymphatic leukemias, congenital fibrosarcoma, mesoblastic nephroma, salivary gland carcinoma, thyroid cancer, malignant melanoma, astrocytoma, and secretory ductal breast carcinoma (Lannon and Sorensen 2005; Skálová et al. 2018; Gatalica et al. 2019).

The *EWSR1* gene in 22q12, which received its name from being the cause of Ewing sarcoma when fused through a t(11;22)(q24;q12) with

FLI1 from 11q24, also not only is capable of recombining with as many as 50 partners, but some of the resulting gene fusions are seen in very different tumors (Fisher 2014). There are even exceptions to the rule that all *EWSR1* rearrangements lead to malignancy; an *EWSR1::POU5F1* fusion gene arising through a t(6;22)(p21;q12) is found both in mucoepidermoid salivary gland carcinoma and benign hidradenoma of the skin (Möller et al. 2008), and fusion of *EWSR1* with *PBX3* (9q33) has been detected in retroperitoneal leiomyoma (Panagopoulos et al. 2015) as well as skeletal and soft tissue myoepitheliomas (Agaram et al. 2015). Evidently, not only may the same gene rearrangement lie behind phenotypically different cancers, but pathogenetically identical rearrangements may occur in both benign and malignant tumors. Conceptually, this does not necessarily constitute any unsurmountable problem as far as thinking about the process of tumorigenesis is concerned – we can always assume that the target cell undergoing transformation was already committed to a certain, narrow differentiation path – but diagnostically it definitely introduces uncertainties of considerable importance.

Where does all this leave us when it comes to phenotypic versus genotypic classifications of the various neoplastic diseases? A suitable bottom line emerging from the above discussion, littered as it is with examples of the shortcomings of both phenotypic and genotypic classifications of neoplastic disease, is that one cannot and should not rely on either principle to the exclusion of the other. Having said this, however, we do believe, or foresee for what it is worth, that the impact of genetic aspects of tumorigenesis is only going to increase further in the years to come. This is because the genetic or genomic (we are of a generation more comfortable with the former word) cell biology parameters are inherently more essential to neoplastic transformation than those operating at the protein or phenotypic level. They are closer to the essence of tumorigenesis or leukemogenesis, so to speak, less likely to represent epiphenomena of little importance. And since we are committed to the view that the deeper our understanding of diseases, especially complex ones like cancer, the more likely it is that somebody – somewhere and some time down the line – can come up with specific remedies that are able to kill the cancer cells while leaving their nonneoplastic neighbors unscathed.

The Ph story constitutes a model example of such *personalized* or *precision medicine* in which intimate knowledge of the two organisms locked in battle, the host and the intruder (the neoplasm), is key to therapeutic success. Ever more sophisticated studies of the leukemic cells – beginning with cytogenetic screening and ending with detailed, molecular-level knowledge of the crucial pathogenetic event – eventually revolutionized the clinical management of patients with Ph-positive leukemias. In other words, the discovery of first the Ph[1], then t(9;22)(q34;q11), then *BCR::ABL1* and its corresponding fusion protein, BCR::ABL1, and finally imatinib and related tyrosine kinase inhibitors is living proof that cancer genetic science works. Pessimism is not justified.

To sum up the clinical aspects of the various schemes according to which neoplastic diseases can be grouped: Although we always strive for ontologically meaningful classifications that tell us important things about pathogenesis, novel classifications may not meet with much enthusiasm from busy clinicians unless they turn out to be clinically useful. If therapeutic interventions become more effective or less harmful within a new classification system, it is embraced, not to mention if new and efficient drugs are also introduced based on the newfound insights into pathogenetic mechanisms that make up the backbone of the new disease categories. Otherwise, new names for old diseases are likely to be met with indifference at best.

This is as it should be, for medicine is a practical more than an academic discipline. At the end of the day, usefulness is king. That cancer cytogenetics continues to provide clinically as well as scientifically useful information is a blessing not only for the patients involved, but also for all who work in the field.

REFERENCES AND FURTHER READING

Agaram, N.P., Chen, H.-W., Zhang, L. et al. (2015). EWSR1-PBX3: a novel gene fusion in myoepithelial tumors. *Genes Chromosomes Cancer* 54: 63–71.

Aplan, P.D. (2006). Chromosomal translocations involving the MLL gene: molecular mechanisms. *DNA Repair* 5: 1265–1272.

Béné, M.C., Grimwade, D., Haferlach, C. et al. (2015). Leukemia diagnosis: today and tomorrow. *Eur. J. Haematol.* 95: 365–373.

Berger, M.F. and Mardis, E.R. (2018). The emerging clinical relevance of genomics in cancer medicine. *Nat. Rev. Clin. Oncol.* 15: 353–365.

Brandal, P., Teixeira, M.R., and Heim, S. (2010). Genotypic and phenotypic classification of cancer: how should the impact of the two diagnostic approaches best be balanced? *Genes Chromosom. Cancer* 49: 763–774.

Fisher, C. (2014). The diversity of soft tissue tumours with EWSR1 gene rearrangements: a review. *Histopathology* 64: 134–150.

Gatalica, Z., Xiu, J., Swensen, J., and Vranic, S. (2019). Molecular characterization of cancers with NTRK gene fusions. *Mod. Pathol.* 32: 147–153.

Greaves, M. (2016). Leukemia 'firsts' in cancer research and treatment. *Nat. Rev. Cancer* 16: 163–172.

Hanahan, D. and Weinberg, R.A. (2011). Hallmarks of cancer: the next generation. *Cell* 144: 646–674.

Harper, D.P. and Aplan, P.D. (2008). Chromosomal rearrangements leading to MLL gene fusions: clinical and biological aspects. *Cancer Res.* 68: 10024–10027.

Heim, S. and Mitelman, F. (2008). Molecular screening for new fusion genes in cancer. *Nat. Genet.* 40: 685–686.

Heim, S. and Mitelman, F. (eds.) (2015). *Cancer Cytogenetics. Chromosomal and Molecular Genetic Aberrations of Tumor Cells*, 4e. Chichester: Wiley-Blackwell.

Kumar-Sinha, C. and Chinnaiyan, A.M. (2018). Precision oncology in the age of integrative genomics. *Nat. Biotechnol.* 36: 46–60.

Lannon, C.L. and Sorensen, P.H. (2005). ETV6-NTRK3: a chimeric protein tyrosine kinase with transformation activity in multiple cell lineages. *Semin. Cancer Biol.* 15: 215–223.

Marschalek, R. (2015). MLL leukemia and future treatment strategies. *Arch. Pharm.* 348: 221–228.

Matsukawa, T. and Aplan, P.D. (2020). Clinical and molecular consequences of fusion genes in myeloid malignancies. *Stem Cells* 38: 1366–1374.

Mertens, F., Johansson, B., Fioretos, T., and Mitelman, F. (2015). The emerging complexity of gene fusions in cancer. *Nat. Rev. Cancer* 15: 371–381.

Möller, E., Stenman, G., Mandahl, N. et al. (2008). POU5F1, encoding a key regulator of stem cell pluripotency, is fused to EWSR1 in hidradenoma of the skin and mucoepidermoid carcinoma of the salivary glands. *J. Pathol.* 215: 78–86.

Panagopoulos, I., Gorunova, L., Bjerkehagen, B., and Heim, S. (2015). Fusion of the genes EWSR1 and PBX3 in retroperitoneal leiomyoma with t(9;22)(q33;q12). *PLoS One* 10: e0124288.

Popper, K.R. (1957). *The Poverty of Historicism*. London: Routledge & Kegan Paul.

Pui, C.H. (2020). Precision medicine in acute lymphoblastic leukemia. *Front. Med.* 14: 689–700.

Rowley, J.D., Le Beau, M.M., and Rabbitts, T.H. (eds.) (2015). *Chromosomal Translocations and Genome Rearrangements in Cancer*. Cham: Springer.

Sandberg, A.A. and Meloni-Ehrig, A.M. (2010). Cytogenetics and genetics of human cancer: methods and accomplishments. *Cancer Genet. Cytogenet.* 203: 102–126.

Sansregret, L., Vanhaesebroeck, B., and Swanton, C. (2018). Determinants and clinical implications of chromosomal instability in cancer. *Nat. Rev. Clin. Oncol.* 15: 139–150.

Skálová, A., Stenman, G., Simpson, R.H.W. et al. (2018). The role of molecular testing in the differential diagnosis of salivary gland carcinomas. *Am. J. Surg. Pathol.* 42: e11–e27.

Studtmann, P. (2008). Aristotle's categories. In: *The Stanford Encyclopedia of Philosophy* (ed. E.N. Zalta). `https://plato.stanford.edu/archives/fall2008/entries/Aristotle-categories/`.

Swerdlow, S.H., Campo, E., Harris, N.L. et al. (eds.) (2017). *WHO Classification of Tumours of Haematopoietic and Lymphoid Tissues*, 4e. Lyon: IARC Press.

WHO (2020). *WHO Classification of Tumours. Soft Tissue and Bone Tumours*, 5e. Lyon: IARC Press.

CHAPTER 15

Where There is Structure, There is Function

In the preceding chapters, we have repeatedly alluded to the extremely important biological dichotomy that separates, and yet simultaneously binds together, genotype and phenotype: The possibilities encoded in our genetic material and the end-result obtained when that same genotype interacts with environmental factors. An even more crucial division – probably the most fundamental and important one there is – exists between structure and function. In theory, both dichotomies come across as clear-cut, so distinctive that it should be easy to determine what is what. In practice, however, this may not always be the case.

In almost all kinds of biological research, emphasis has over the last hundred years or so gradually switched from studies of structure to attempts at finding out how things work or, in the field of pathology, how things do *not* function properly so that diseases, including neoplasms, occur causing ill-being and death. We have often noticed this changing attitude in review committees where grant proposals are evaluated; criticism that an application is "merely descriptive" is frequently leveled against morphology-oriented research. Consequently, unless the scientific plan under scrutiny makes an active, and sometimes less than truthful, claim that a study of function is one of the aims, funding is seldom granted. Not only should the research employ "cutting-edge molecular genetic methodology," whatever that may mean this year or even this month, in order to end up among the select few looked upon with mercy, but the effect(s) of whatever

Abnormal Chromosomes: The Past, Present, and Future of Cancer Cytogenetics.
Sverre Heim and Felix Mitelman.

gene-level aberrations are examined should be subjected to assessment at the protein level or even in intact organisms. In an ideal world, this is all fine; it is always an ultimate aim to undertake the most comprehensive, even complete, investigation possible. In reality, however, not rarely and for a variety of reasons one has to accept something considerably less and not so impressive. Here as elsewhere, the best should not be allowed to become an enemy of the good.

Chromosomes are obviously structures that are part of the genome, inasmuch as they contain or carry the physical threads on which the genetic code is imprinted in the form of base triplets, but it is equally possible, even warranted, to view them as phenotypic manifestations. In that sense, one may claim, tongue in cheek, that they are both the cause (genotype) and the result (phenotype) of themselves. Such a view on the matter does not exactly make things easier as far as clarification of their correct position in the greater scheme of things is concerned, but it does put the finger on the duplicity of chromosomes as both structural and functional elements of importance for how cells and organisms work. Some relationships *are* inherently complicated – no truthful person ever promised otherwise. Solutions, decisions or explanations that do not pay due attention to this aspect of reality are oversimplifications; they are never helpful when it comes to understanding how things work.

Once the genetic code was shown to consist of base triplets neatly strung out as codons along a protein backbone, research interest grew rapidly as to the role of genes in health and disease, including how acquired genetic changes drive neoplastic transformation. As a consequence, even within the paradigm of Boveri's somatic mutation theory, researchers increasingly began to focus on the role of primary DNA structure alterations that occur when a formerly normal cell undergoes neoplastic transformation, rather than which chromosomal aberrations accompany or even cause the gene-level changes. But was this the final say in the matter? Is it so that genes are all there is to heredity at both the organism and somatic cell level, that no chromosome-level functions exist that cannot be accounted for by a detailed study of the DNA's primary structure?

We refuse to believe that this is the case. When attempting to argue in favor of a role for additional, nongenic modes of asserting regulatory control over the genome, it is reasonable to introduce the terms *epigenetics* and *position effect*. They refer to phenomena that are generally crucial not only in our understanding of how genomic control of biological processes is exercised, but perhaps also more specifically to the question of how chromosomal structures influence function in other ways than through classic gene action.

Position effect refers to the observation that expression of a gene may be altered when its location on a chromosome is changed, including by translocation or some other exchange of chromosomal material. The phenomenon was studied in particular detail in *Drosophila* where eye color may depend on whether a color gene is closer or more distant to chromosomal heterochromatin. Also Barbara McClintock's seminal – and for decades neglected – studies of how the position of genes can alter the phenotype in maize should be mentioned in the same vein. All of a sudden, the stable "beads on a string" model of genes along chromosome arms was shown to be very much wanting and words like "transposons" and "jumping genes" entered the core vocabulary of genetics.

A special position effect is well known from cancer cytogenetics, namely the recombination of *MYC* (or another oncogene prone to undergo quantitative activation) with immunoglobulin genes (or other loci under constitutive stimulation by enhancers or other strong regulatory elements); as a consequence, increased and untimely production of an otherwise normal oncoprotein takes place, leading to neoplastic transformation (see Chapter 8). Though first detected in Burkitt lymphoma and related leukemias, similar rearrangements are found also in other neoplastic conditions, often as a main pathogenetic pathway.

One may protest the relevance of this example inasmuch as the mechanisms of altered expression of *MYC* are known in molecular detail. Therefore, nothing supragenic can or should be read into it; an oncogene comes under the influence of another gene's regulatory elements and that is basically that. But what if something similar (similar in the sense that the end-result is again altered gene expression, but nevertheless different considering that the regulatory change is mediated via a mechanism not involving any change in DNA primary structure) also occurs, perhaps even more frequently, and in both health and neoplastic disease? We choose to reason – or speculate, if you prefer – about this possibility from a cytogenetic or chromosomal angle which may be said to belong under the overall umbrella of epigenetics, the study of heritable changes of phenotype brought about by something other than DNA sequence alterations.

Consider first the common cytogenetic case of nonrandomly acquired numerical chromosome aberrations, especially gains of one or more chromosomes that look completely normal from a structural point of view, say trisomy 8 in acute myeloid leukemia or the massive hyperdiploidy genomic pattern so typical of childhood acute lymphoblastic leukemia. Without any doubt, numerous quantitative expression studies have borne this out; these aberrations do not correspond functionally to any simple dosage effect of the genes now present in three (or four for tetrasomies) copies. The same seems to apply generally to copy number variations in neoplastic cells; you cannot *a priori* conclude that you have 150% gene expression just because you have three copies of an allele. For many cancers and leukemias, practice has shown that no simple, linear correlation exists between the number of gene copies or alleles present in a genome and the output of that gene's product. Clearly, expression is controlled at another, we assume higher, organizational level within the cell nucleus, and about this supragenic control system preciously little knowledge is available.

Maybe it is unconventional to use the term *epigenetic* about such putative higher-order expression control, but we cannot see it as misplaced for such stable phenotypic changes caused by something other than DNA sequence alterations. Or perhaps a better word is *epigenic*, for our whole point is that although, without a shadow of doubt, *much* genetic control is exerted at the level of base pair sequences in genes and their regulatory elements, it is entirely possible, we think likely, that *some* control depends on structures and mechanisms that emanate from much larger structures. It may also be appropriate to refer to this type of expression regulation as *emergent* since it manifests at a greater complexity level. If so, it comes across as a cellular process with functional consequences that cannot be foreseen based on knowledge of the cell's DNA primary structure. By way of a classic out-of-area illustration, there is no way one can deduce the existence of the English language from knowledge that such a thing exists as the Latin alphabet, just as one cannot surmise the existence of Shakespeare's plays just because there are people who speak and write English. Whereas reductionist thinking and mechanisms (expression control is in its entirety dependent on DNA- and gene-level alterations and stimuli) thus certainly account for part of the orchestration of which genes work and which do not at any given time, there are, in our antireductionist opinion, other situations when this is not so. In short, gene control takes place at several levels of intranucleic organization and complexity, of which chromosomal structures are probably one, possibly even the most important.

When a gene locus is moved to another position on the same or another chromosome, therefore, this may affect its expression by upregulating or downregulating it, and something broadly similar appears to go also for loci on supernumerary chromosomes. If the whole chromosome complement is in upheaval, furthermore, for instance when, as part of neoplastic transformation, massive numerical and/or structural chromosome aberrations are acquired (Figure 15.1), we can safely assume that the intranucleic position of

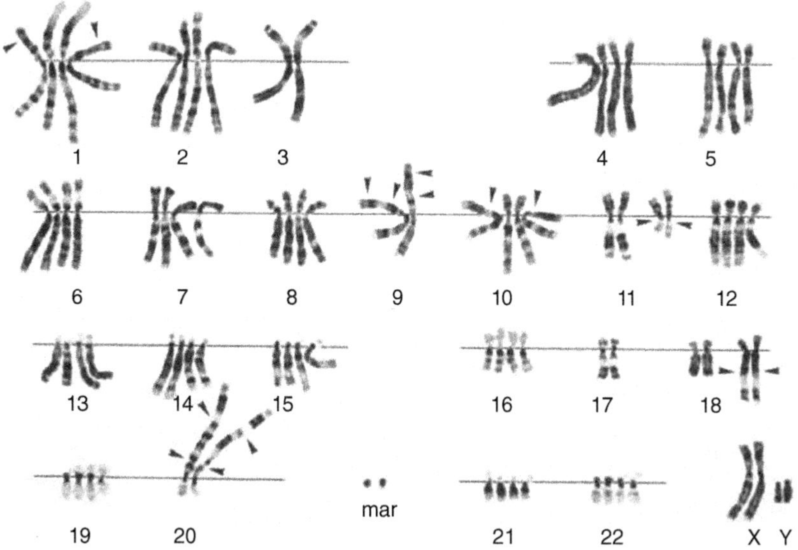

FIGURE 15.1 Karyogram showing massive clonal numerical (including endoreduplication) and structural chromosome aberrations in a cell cultured from a duodenal adenocarcinoma (Gorunova et al. 1995). There is every reason to assume that, in addition to transcription changes caused by alteration of DNA primary structure, regulatory changes at several supragenic organization levels are also operative. Source: Reproduced with permission from Elsevier.

"working" chromosomes and chromosome segments is disturbed. They may find themselves stranded in new domains under new regulatory influences, a repositioning that results in a novel expression profile of the malignant cell. We do not know how this type of domain regulation is accomplished or maintained during health and disease – in that sense, our hypothesis certainly deserves to be branded as speculative – but it would be strange if it were not there.

The shape, content, and position of chromosomes together constitute one distinct and obvious level of genetic complexity and organization. There is a long history of their microscopic study and a much shorter history of knowledge as to how chromosomes carry and regulate genes. Both investigative approaches are going to be continued, of course always using the technologies deemed best suited to the purpose. One can only hope that a proper balance is struck between studies of structure and function when tomorrow's researchers go about that task, whether they work from a chemical or morphological starting point. If not, both strategies will suffer, for structure and function are intimately intertwined; any study of one without a constant eye on the other is doomed to be a lopsided enterprise.

It may sound paradoxical, quaint even, but there is more to studies of chromosomes than meets the eye. There is more we do not know about their importance and behavior, perhaps even their structure, than has already been discovered.

From structure springs function, both inside and outside cells. The future of cytogenetics may be brighter than most workers within the field currently imagine, especially for those who are simultaneously able to glance in the side mirrors at function as they drive down discovery lane.

REFERENCES AND FURTHER READING

Akdemir, K.C., Le, V.T., Kim, J.M. et al. (2020). Somatic mutation distributions in cancer genomes vary with three-dimensional chromatin structure. *Nat. Genet.* 52: 1178–1188.

Bakhoum, S.F. and Cantley, L.C. (2018). The multifaceted role of chromosomal instability in cancer and its microenvironment. *Cell* 174: 1347–1360.

Ben-David, U. and Amon, A. (2020). Context is everything: aneuploidy in cancer. *Nat. Rev. Genet.* 21: 44–62.

Bird, A. (2007). Perceptions of epigenetics. *Nature* 447: 396–398.

Bustin, M. and Misteli, T. (2016). Nongenetic functions of the genome. *Science* 352: aad6933.

Cavalli, G. and Heard, E. (2019). Advances in epigenetics link genetics to the environment and disease. *Nature* 571: 489–499.

Chatterjee, A., Rodger, E.J., and Eccles, M.R. (2018). Epigenetic drivers of tumourigenesis and cancer metastasis. *Semin. Cancer Biol.* 51: 149–159.

Chunduri, N.K. and Storchová, Z. (2019). The diverse consequences of aneuploidy. *Nat. Cell Biol.* 21: 54–62.

Cremer, M. and Cremer, T. (2019). Nuclear compartmentalization, dynamics, and function of regulatory DNA sequences. *Genes Chromosomes Cancer* 58: 427–436.

Fedoroff, N.V. (2012). McClintock's challenge in the 21st century. *Proc. Natl Acad. Sci. USA* 109: 20200–20203.

Gorunova, L., Johansson, B., Dawiskiba, S. et al. (1995). Cytogenetically detected clonal heterogeneity in a duodenal adenocarcinoma. *Cancer Genet. Cytogenet.* 82: 146–150.

Hanahan, D. and Weinberg, R.A. (2011). Hallmarks of cancer: the next generation. *Cell* 144: 646–674.

Kneissig, M., Bernhard, S., and Storchova, Z. (2019). Modelling chromosome structural and copy number changes to understand cancer genomes. *Curr. Opin. Genet. Dev.* 54: 25–32.

Misteli, T. (2007). Beyond the sequence: cellular organization of genome function. *Cell* 128: 787–800.

Ravindran, S. (2012). Barbara McClintock and the discovery of jumping genes. *Proc. Natl Acad. Sci. USA* 109: 20198–20199.

Uhlen, M., Zhang, C., Lee, S. et al. (2017). A pathology atlas of the human cancer transcriptome. *Science* 357: eaan2507.

Weiler, K. and Wakimoto, B. (1995). Heterochromatin and gene expression in drosophila. *Annu. Rev. Genet.* 29: 577–605.

CHAPTER 16

Which Resolution Level is Optimally Suited to Answer Which Questions? Seeing Never Goes Out of Fashion . . .

When undertaking to study a complex structure – and humans are about as intricate and complex as anything can be whether looked at during health or unhealth – a primary concern is to decide at which resolution level the examination should be conducted. Is it more advantageous to ask questions that are meaningful and can be answered at some composite, higher level of organization, i.e., to focus on relatively big components, or should one always attempt to probe as deeply as possible into the smallest constituents, which for genetic disorders would be genes or even the base pair primary structure they carry? We have touched upon this investigative dichotomy repeatedly in preceding chapters, for it is truly fundamental to all thinking about the role of cytogenetics in cancer research. It should surprise nobody that the same principal difficulty or choice will face future examiners of the acquired genomic alterations that lie behind neoplastic processes.

Let us by way of two indirect examples emphasize the empirical fact that examination at the largest possible resolution level is not always optimal

Abnormal Chromosomes: The Past, Present, and Future of Cancer Cytogenetics.
Sverre Heim and Felix Mitelman.

when striving to obtain an understanding of disease processes. When we were still young, the environmental damage done to Europe's dwindling forests by acid rain stemming from industrial pollution (largely nitric and sulfuric acids) received a lot of attention, and probably rightly so. How can one best assess such damage in any particular forest? By a variety of methods, of course, both morphological and chemical, and possibly others as well. But if we restrain our attempts to answers involving inspection, which surely is key to any morphological assessment, at which organization level should examiners focus their efforts? Some good may possibly come from microscopic examination of representative leaves, maybe even electronic microscopic studies of subcellular structures for all we know, but probably more relevant information could be obtained – faster, cheaper, and more reliably – by having a competent woodsman, possibly together with a specialist in dendropathology, stroll through the forest and take a good look. Perhaps aerial photography of the region might be an excellent method in order to, literally, get the best overall picture of a forest's condition, for a forest is surely more than just a collection of trees. At any rate, and we repeat ourselves but do so knowing that the point is all too often forgotten: The type of knowledge one is after should decide the investigative methods one opts for.

Since we mentioned the possibility of using electron microscopy in the acid rain scenario above, it may be appropriate to look more carefully at the historic relevance of this investigative technique in a medical, including oncological, setting. Fantastic though electron microscopy is, both in an esthetic sense (is there anything more beautiful than three-dimensional scanning EM pictures of exceedingly small structures?) and as a means to visualize what is too little to be seen by light microscopy, the technology's actual usefulness in cancer pathology has proven to be disappointingly limited. The point we want to make is that sometimes, seeing too many details obscures one's ability to assess the truly informative changes that characterize a tissue, a lesion, or a cell involved in a particular disease process. Small may well be beautiful, but this does not always mean that smaller is better, neither for diagnostic nor scientific purposes.

This conclusion definitely holds also for diseases brought about by acquired changes of the cells' genome. Some information about the nature of a given neoplastic process is best obtained at the tissue level, some is obtainable by looking at cells, nuclei or chromosomes, while other changes have to be analyzed at the level of genes and primary DNA structure for us to be able to make sense of what is going on.

Before moving on to a discussion of the joint cytogenetic-molecular genetic approach we see as optimal in many studies of cancer cells, we would like to relay a little example of how karyotyping remains, and is likely to remain for the foreseeable future, necessary to the detection of some characteristic genomic aberration patterns in distinctive malignant disorders. It literally stems from last week's diagnostic work and illustrates the type of information obtainable only – or almost only, one should never be too certain regarding future technological developments – by direct chromosome banding analysis, and yet the outcome is of considerable both diagnostic and scientific importance.

A kidney cancer sample had been sent for chromosome analysis. Several aberrations were found corresponding to two related clones (Figure 16.1), one the duplicate of the other. The simplest, hypodiploid one showed losses of the Y chromosome and one copy of each of chromosomes 1, 2, 6, 10, 13, 17, and 21. Its more numerous duplicate had disomies for the above-mentioned autosomes and X (the Y was still lost), but tetrasomies for the remainder, exactly as should be if the first clone had undergone reduplication. No structural abnormalities were seen. Hence, without prior knowledge that the hypodiploid clone existed, only a strange pattern of multiple acquired tetrasomies would have met the eye, and nobody could have guessed, let alone concluded, that the seeming disomies were actually duplicated monosomies. Stranger misinterpretations may and have happened if one relies exclusively on, for example, array comparative genomic hybridization (CGH), possibly supplemented with something even more modern such as single nucleotide polymorphism (SNP) arrays, as a means to detect genomic imbalances. In the case above, consultation of the relevant literature quickly reminded us that the monosomies defining the original clone are indeed known in cancer cytogenetics, namely as the specific aberration pattern of chromophobe renal cell carcinomas (Kovacs and Kovacs 1992; Gunawan et al. 1999; Mertz et al. 2008), a malignancy that accounts for only 5% of all kidney cancers.

There are several lessons to be learned from this little diagnostic story which, by the way, is not as exotic as it may seem to those who do not work daily in cancer cytogenetics. Furthermore, several of the take-home lessons are likely to be equally informative about things to come as they were or are about yesterday's and today's practices.

The most important is that many characteristic chromosomal aberration patterns exist in both common and rare neoplasms. Even in the most extensively studied malignancies, the leukemias, it is not uncommon to come across something that has not been seen before, or at least was not reported before. Lots of potentially important information is there for the picking, often right before our eyes, only somebody has to perform the analyses and

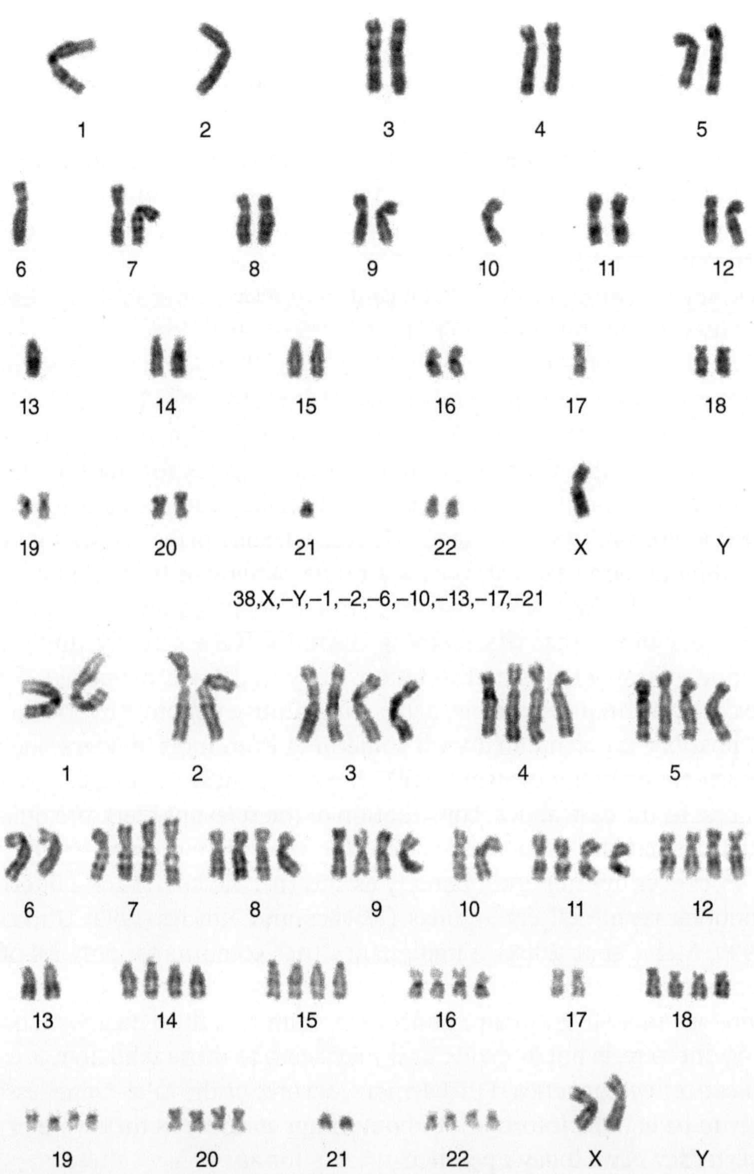

76,XX,–Y,–1,–2,+3,+4,+5,–6,+7,+8,+9,–10,+11,+12,–13,+14,+15,+16,–17,+18,+19,+20,–21,+22

FIGURE 16.1 Two clones detected by G-banding analysis of cells cultured from a chromophobe renal cell carcinoma. All chromosome aberrations were numerical. One clone (top karyogram) was characterized by multiple monosomies stemming from losses of the Y and one copy of each of chromosomes 1, 2, 6, 10, 13, 17, and 21. The second, related clone (bottom karyogram) had arisen through duplication of the first. Hence, its apparent aberrations compared with the normal 46,XY pattern were loss of Y with gain of one X together with tetrasomies for autosomes 3, 4, 5, 7, 8, 9, 11, 12, 14, 15, 16, 18, 19, 20, and 22 (in the legend, however, the description is relative to a normal triploid karyotype, 69,XXY, because the actual chromosome number is closer to 3n than 2n). Such aberration patterns cannot, at least not easily, be detected by any other techniques than direct chromosome analysis. Though most complex, disease-specific karyotypic deviations characterized by numerical aberrations only are seen in rare types of neoplasia, many such correlations are known to exist and probably additional examples would be detected if cytogenetic analyses were performed more consistently whenever tumors are diagnosed. In general, and especially if we lump together aberration patterns that may include both structural and numerical chromosomal changes, the one karyotypic subgroup that comes across as most numerous is probably the one containing all the rare but specific (or at least characteristic) aberration patterns. This fact alone is a strong argument to commence all genetic examinations of neoplastic cells using a technique such as G-banding that has screening qualities.

understand that scientific serendipity is smiling at them when they find something rare or hitherto undescribed. Within every new case may lurk something functionally, diagnostically or prognostically important, but it has to be recognized as such by at least one person in the team performing the relevant diagnostic work, and that someone has to have the necessary energy to make his/her findings public. And rest assured: Once a strange or unique aberration pattern has been described, experience tells that there are always others who have seen a similar case, or you will come across another one in your own practice shortly afterwards. Unique cases thus turn into rare cases and what is rare to begin with ultimately becomes established as yet another perhaps not so frequent, but eventually well-established clinical-cytogenetic relationship. *Only if workers in the field cease to perform the analyses and publish unusual findings will the accumulation of new knowledge of this kind grind to a halt.* And this would be truly sad, for as we have already argued, where there is structure, there is function. We may not know today how multiple chromosome losses, or gains for that matter, act pathogenetically, but their specific occurrence in a type of cancer does tell us that something important has occurred.

That the described case was characterized by the presence of only numerical chromosome aberrations was not coincidental given our chosen context, for these changes are the most difficult ones to detect and interpret by molecular genetic technologies, including the many transitional techniques that occupy the border zone toward banding cytogenetics. As already pointed out, neither CGH-based nor SNP arrays can detect the coexistence of two such clones, one the duplicate of the other, in any straightforward and reliable manner. Also state-of-the-art sequencing techniques are helpless in this regard, at least unless very indirect and intricate modifications constitute part of the investigative procedure. To those who opine that aberration patterns of the kind related above probably do not tell an informative story about carcinogenesis anyway, and we have heard this "argument" often, we would like to counter: How on earth can anybody know anything about such likelihoods? What is certain is that a finding (the strange numerical aberration pattern) was made that represents a strict and unforeseen correlation within cancer biology. Such deviations from the random are strong indicators that something functionally significant is going on, and we most emphatically advocate the view that to disregard them – or even worse, not to look for such genomic imbalances at all or do so only using inappropriate techniques – would be to sin against the fundamental principles of scientific research. The Devil or the accident-prone, stochastic nature of cell replication or whatever has seen to it that "experiments" abound of how the well-balanced interplay among bodily cells is taken over by aggressive, genetically altered clones, and the least we can do is to try our utmost to understand how things went wrong.

In recent years, most emphasis in cancer cytogenetics has been on the detection of fusion genes generated by structural chromosomal rearrangements (Mertens et al. 2015). It is within this field that the impact of ever more modern molecular genetic techniques has been the greatest. Probably this will be an even more distinctive feature of the future, so the following question therefore has to be asked: Does not the introduction of new sequencing techniques into the genetic analysis of neoplasia make analyses by banding cytogenetics obsolete, at least as far as the hunt for oncogenic fusion genes is concerned? Indeed, at least since the early 1980s the imminent demise of cytogenetics as a viable technique has been proclaimed by ever-eager advocates of a tomorrow that is always said to be just around the corner.

Let us first make clear that the various massively parallel DNA- and RNA-based sequencing (MPS) techniques (Metzker 2010; Slatko et al. 2018; McCombie et al. 2019) introduced into genetic research and diagnostics in recent years – whole-genome sequencing (WGS), whole-exome sequencing (WES), and RNA sequencing (RNA-seq) – obviously represent truly wonderful developments. They collectively constitute huge steps forward in our abilities to sort out gene-level genomic alterations in constitutional as well as somatic cell genetics. Sequencing has revolutionized the search for oncogenic mutations, including gene fusions (Gao et al. 2018; Hu et al. 2018; Alexandrov et al. 2020; Calabrese et al. 2020) in many types of neoplasia. So tremendous has the impact been that one almost feels like forgiving the many scientific salesmen who present every technologically novel twist as an improvement almost beyond belief.

No technique is perfect, however, not even MPS, something that comes as no surprise to anyone with an active interest in history, including the history of science. RNA-seq or MPS analyses of tumor cells often reveal hundreds of candidate fusion genes, all or almost all of which, upon painstaking validation using combinations of fluorescence *in situ* hybridization (FISH), Sanger sequencing, and other molecular methods, turn out to be false leads. Thus, not only are there numerous examples of "false negatives" whenever MPS is used as a stand-alone technique to find the putative decisive transformation-inducing fusion gene of a given neoplasm, but often you end up with a large number of technical, biological, and, perhaps in particular, clinical "false positives." This inevitably makes the assessment of which fusions are important and which are noise extremely difficult (Panagopoulos et al. 2014; Johansson et al. 2019).

How, then, can the number of candidate genes be reduced to something more manageable, perhaps at the same time gaining an independent hint as to where the important gene(s) might reside in the genome? With an alternative plan to weed out the less likely candidates before full-scale verification/falsification measures are implemented, the remaining, many fewer recombinants of interest could then be scrutinized more economically, both money-wise and in terms of the labor required. This would be done by more accurate, but also more time-consuming, methods than those involving computational searches for fusion genes by means of FusionFinder, FusionMap, nFuse, FusionSeq, and other types of software (Latysheva and Babu 2016).

One way to go about this, which to some extent combines the old candidate gene approach with the brute force methods alluded to above, involves sequential analyses of neoplastic cells by first karyotyping, then RNA sequencing (Panagopoulos et al. 2014). This, of course, is not meaningful in those instances where no cytogenetic rearrangement is found hinting that a fusion gene may have been generated, but then again, enough cases exist, also in highly malignant, epithelial tumors, that do have such chromosomal features. You may visualize the two investigative approaches as providing independent information grids that can be laid on top of one another across the entire genomic landscape; every cytogenetic rearrangement represents a point of interest, as does every candidate fusion gene marked by RNA-seq.

Where the point(s) of both grids coincide, one lying on top of the other, a position has been identified where it is likely that a pathogenetically important, oncogenic fusion gene was generated during tumorigenesis. This candidate hybrid, first identified by RNA-seq, must now have its actual existence verified by other techniques, typically a combination of FISH and Sanger sequencing. However, this still rather cumbersome verification/falsification procedure can now be concentrated on only one or a few genes, not the 100+ candidates that would all have had the same *a priori* likelihood of being biologically important if no cytogenetic data were available. The fewer uninteresting trees one has to bark up, the better.

Many pathogenetically important genes have been identified using this approach which significantly reduces the amount of work needed to probe in a meaningful manner the oncogenic mechanisms leading to any given neoplasm. True, tumors that do not have any characteristic chromosomal aberrations cannot be examined using this approach, but the circumstance that not *all* cases can be evaluated satisfactorily using any given technique or combination of techniques is hardly new in medicine. One does the best one can, in science as elsewhere. Here again, we face a situation in which it is advisable not to turn the best into an enemy of the good.

All technologies change over time, no matter which principle they build on. Diagnostic as well as research work depends on a suitable balance between rationality and empirical efforts; it is a multipronged approach in which visualization of the object or phenomenon under scrutiny always contributes mightily to our attempts to understand what is going on. No matter what happens to classic light microscopy, seeing is never going to be completely replaced as one of the important modes of investigation, no matter how many other, more indirect investigative technologies are introduced. In addition, new gadgets are constantly being invented to improve our microscopic eyesight as well, so visualization, too, is far from having reached any evolutionary final stop.

An old proverb maintains that "seeing is believing," and at least to some extent we find it spot on. Seeing chromosomes and the genes they harbor, in one way or another, is part of reasoned examination of many disorders that involve an altered genomic constitution of somatic cells, including cancer, so seeing is also understanding.

To the best of our understanding, therefore, no good reason exists to believe that seeing can ever satisfactorily be substituted by other modes of data gathering with simultaneous processing into an understanding of what one observes, nor is it reasonable to strive for anything so unprecedented in medicine or research. Microscopic analyses of relevant structures are unbiased, cheap, and robust, and they often lead to the gain of highly relevant information. The best results are typically obtained when combinations of techniques are brought to bear, however. This, too, is known to reflect a reasonable, balanced best-by-test attitude throughout the history of medical progress, testimony to the fact that also in this respect the past and the future are tightly interconnected.

REFERENCES AND FURTHER READING

Alexandrov, L.B., Kim, J., Haradhvala, N.J. et al. (2020). The repertoire of mutational signatures in human cancer. *Nature* 578: 94–101.

Bacher, U., Shumilov, E., Flach, J. et al. (2018). Challenges in the introduction of next-generation sequencing (NGS) for diagnostics of myeloid malignancies into clinical routine use. *Blood Cancer J.* 8: 113.

Calabrese, C., Davidson, N.R., Demircioglu, D. et al. (2020). Genomic basis for RNA alterations in cancer. *Nature* 578: 129–136.

Dal Cin, P., Qian, X., and Cibas, E.S. (2013). The marriage of cytology and cytogenetics. *Cancer Cytopathol.* 121: 279–290.

Gao, Q., Liang, W.W., Foltz, S.M. et al. (2018). Driver fusions and their implications in the development and treatment of human cancers. *Cell Rep.* 23: 227–238.e3.

Gunawan, B., Bergmann, F., Braun, S. et al. (1999). Polyploidization and losses of chromosomes 1, 2, 6, 10, 13, and 17 in three cases of chromophobe renal cell carcinomas. *Cancer Genet. Cytogenet.* 110: 57–61.

Hochstenbach, R., Liehr, T., and Hastings, R.J. (2021). Chromosomes in the genomic age. Preserving cytogenomic competence of diagnostic genome laboratories. *Eur. J. Hum. Genet.* 29: 541–552.

Hu, X., Wang, Q., Tang, M. et al. (2018). TumorFusions: an integrative resource for cancer-associated transcript fusions. *Nucleic Acids Res.* 46: D1144–D1149.

Johansson, B., Mertens, F., Schyman, T. et al. (2019). Most gene fusions in cancer are stochastic events. *Genes Chromosomes Cancer* 58: 607–611.

Kovacs, A. and Kovacs, G. (1992). Low chromosome number in chromophobe renal cell carcinomas. *Genes Chromosomes Cancer* 4: 267–268.

Latysheva, N.S. and Babu, M.M. (2016). Discovering and understanding oncogenic gene fusions through data intensive computational approaches. *Nucleic Acids Res.* 44: 4487–4503.

Lazarevic, V.L. and Johansson, B. (2020). Why classical cytogenetics still matters in acute myeloid leukemia. *Expert. Rev. Hematol.* 13: 95–97.

Mareschal, S., Palau, A., Lindberg, J. et al. (2021). Challenging conventional karyotyping by next-generation karyotyping in 281 intensively treated patients with AML. *Blood Adv.* 5: 1003–1016.

McCombie, W.R., McPherson, J.D., and Mardis, E.R. (2019). Next-generation sequencing technologies. *Cold Spring Harb. Perspect. Med.* 9 (11): a036798.

Mertens, F., Johansson, B., Fioretos, T., and Mitelman, F. (2015). The emerging complexity of gene fusions in cancer. *Nat. Rev. Cancer* 15: 371–381.

Mertz, K.D., Demichelis, F., Sboner, A. et al. (2008). Association of cytokeratin 7 and 19 expression with genomic stability and favorable prognosis in clear cell renal cell cancer. *Int. J. Cancer* 123: 569–576.

Metzker, M.L. (2010). Sequencing technologies the next generation. *Nat. Rev. Genet.* 11: 31–46.

Misteli, T. (2007). Beyond the sequence: cellular organization of genome function. *Cell* 128: 787–800.

Mitelman, F., Johansson, B., and Mertens, F. (2007). The impact of translocations and gene fusions on cancer causation. *Nat. Rev. Cancer* 7: 233–245.

Panagopoulos, I., Thorsen, J., Gorunova, L. et al. (2014). Sequential combination of karyotyping and RNA-sequencing in the search for cancer-specific fusion genes. *Int. J. Biochem. Cell Biol.* 53: 462–465.

Rack, K.A., van den Berg, E., Haferlach, C. et al. (2019). European recommendations and quality assurance for cytogenomic analysis of haematological neoplasms. *Leukemia* 33: 1851–1867.

Rosai, J. (2007). Why microscopy will remain a cornerstone of surgical pathology. *Lab. Investig.* 87: 403–408.

Slatko, B.E., Gardner, A.F., and Ausubel, F.M. (2018). Overview of next-generation sequencing technologies. *Curr. Protoc. Mol. Biol.* 22: e59.

Wohlleben, P. (2015). *The Hidden Life of Trees: What they Feel, How They Communicate – Discoveries from a Secret World.* Vancouver: Greystone Books.

CHAPTER 17

Are New Technical Breakthroughs on the Horizon?

New methods to analyze the genomic constitution of disease lesions, even of individual cells, are going to be introduced in the future, that much is certain, at least if we humans are not stupid enough to destroy ourselves and thereby sever the continuous development of technology and reasoned thinking that characterized our past. Whether such developments are close enough in time to be depicted as being "on the horizon," and massive enough to deserve being thought of as "breakthroughs," is a moot point. We cannot know, but we would like to take this opportunity to indulge in a bit of prophesying, or wishful thinking if you like, about what may come to pass. So as not to end up in a total wilderness of mostly unrealistic hopes and dreams, we shall in the main restrict ourselves to one small – but cytogenetically important! – "tactical" improvement we think is long overdue, and another "strategic" shift in emphasis *cum* improvement that, if it ever materializes, would revolutionize cytogenetics as we now know it.

Since today's knowledge about cell biology is vastly improved compared with what we learned back in the day, one would imagine that researchers' ability to induce *in vitro* mitotic division in this or that group of cultured cells should no longer constitute any major challenge. And yet, whereas phytohemagglutinin (PHA) has been used for more than half a century to stimulate cell division in blood cultures – it is the T-lineage lymphocytes

Sverre Heim and Felix Mitelman.

that divide – similar selective mitosis induction methods are not available for the parenchyma cells of the wide array of neoplastic processes we examine cytogenetically for clinical or research purposes. In both suspension cultures and when culturing cells from solid tumors on a matrix, one never knows beforehand whether stromal elements, cells belonging to the neoplastic parenchyma or some other cells that became part of the sample through whatever mechanism, are the more likely to begin dividing. Sometimes chromosomally abnormal leukemic cells outgrow their normal neighbors, but sometimes it is the other way round. Often stromal cells take over a solid tumor culture if it remains unharvested for weeks, but on other occasions it is the truly neoplastic tumor cells that are late in entering mitosis. There is no distinct pattern to discern, and the diagnostic usefulness of cancer cytogenetic analyses most certainly suffers from this chronic difficulty and uncertainty.

What we as cancer cytogeneticists have long hoped for from our cell biologist colleagues is some kind of specific "golden dust" that could induce mitoses in neoplastic cells exclusively or, at the very least, preferentially. That would mean that we were able to examine cells with abnormal chromosomes in a higher percentage of cases, but also the eternal problem of how to interpret karyotypically normal cultures would finally become less common and disturbing. It is worth underscoring that there is nothing sophisticated about such a wish to be able to look only at the cells of interest in our search for acquired chromosome abnormalities, we even admit to it being rather naive, but it really would be such a wonderful improvement.

Our second, strategic wish for the future of cytogenetics is a lot more complicated than finding something that selectively induces cell division. We shall enter the theme via an attempt undertaken in the 1980s to visualize both cellular karyotype and phenotype.

Knuutila and coworkers in Helsinki, Finland, at that time described (Teerenhovi et al. 1984; Knuutila et al. 1994) a manual for how to analyze simultaneously or sequentially cell morphology, immunophenotype, and chromosome constitution. Their ambitious MAC method – for Morphology, Antibodies, and Chromosomes – attempted to bridge the seemingly eternal analytical gap that has plagued all cytological examinations since the very beginning: You can either obtain a decent picture of a given cell's genotype (chromosomes) or its phenotype (antibodies and morphology in the MAC perspective), but not both at the same time. Since our entrance into the difficulties this entails is through a cytogenetic route – the study of chromosomes – this means that we cannot directly know anything about the phenotypic features of a cell whose mitotic chromosomes we inspect, for all cytoplasmic structures were removed during the process of making suitable chromosome preparations. By the way, this is another reason why we argued so strongly above for the necessity of obtaining new and truly selective mitogens, substances capable of inducing mitotic activity in only one or a few cell types, not others. As the situation is now, we conclude indirectly that whenever clonal aberrations are detected, then the examined cells must be part of the neoplastic parenchyma, whereas we have no direct certainty as to which tumor component those cells belong that show a normal karyotype.

Although these pioneering efforts were subsequently followed up by several modifications and improvements, the MAC approach never met with much success. Apparently, the methodological compromises necessary to arrive at both a phenotypic and a karyotypic picture of the cells under scrutiny were too taxing. As it was, the chromosome analyses possible on preparations containing dividing cells that had retained at least part of their outer membrane were exceedingly coarse in the sense that only gross abnormalities could be seen, and the same applied to the distribution of surface antibodies and other extranuclear morphological features.

At about the same time, in 1992, Weber-Matthiesen and coworkers developed a technique combining fluorescence in situ hybridization (FISH) with fluorescence immunophenotyping. This FICTION (**f**luorescence **i**mmunophenotyping and interphase **c**ytogenetics as a **t**ool for the **i**nvestigation **o**f **n**eoplasms) method also allowed the simultaneous study of morphological, immunophenotypic, and genetic features of single cells, and more recently a modified multicolor FICTION technology was introduced by Siebert and coworkers (Martin-Subero et al. 2002; Giefing and Siebert 2019). These techniques have so far mainly been used to examine hematological disorders, in particular malignant lymphomas, but of course one hopes that, upon further improvements, they will one fine day find a place in solid tumor cytology. Given that so much research is being conducted along these lines, combining sophisticated new instruments with molecular technologies, the goal of obtaining at some future point a more detailed, combined genotypic-phenotypic picture of neoplastic cells does not seem entirely unrealistic.

Flow cytometry has been used for years to enable cytogenetic studies of phenotypically defined cell populations, with new procedures steadily being introduced to sort labeled and tagged cells in order to identify both phenotypic and genotypic traits of interest. Combined with digital microscopy, such refined flow cytometric analyses also provide a powerful tool to look for cytogenetic changes in cells whose phenotypic identity is already known. The ability to visualize the cells of interest, to determine the pattern and localization of expressed antigens, and to assess the presence or absence of defined cytogenetic abnormalities in one integrated, automated high-throughput test relying on imaging flow cytometry has even been presented as a new paradigm for cytogenetic analyses of cancer cells (Grimwade et al. 2017). Perhaps these authors are right. At the very least, the methodologies involved are truly fascinating.

Although cytology and cytogenetics have traditionally been the only methods at hand to assess the genetic composition of individual cells, powerful molecular genetic methodologies are now bridging the gap between the worlds of chemistry and microscopy. In particular, single-cell RNA sequencing, whereby information is obtained about the cellular phenotype, is increasingly being combined with the simultaneous detection of gene-level markers (Dey et al. 2015; Suvá and Tirosh 2019; Zhou et al. 2020). The speed of technical innovations in this area of research is truly breathtaking. Claims are being made that soon a complete genetic profile of clones or other groups of cells, including an overview of each cell's copy number changes and specific gene rearrangements, will be obtainable using the new sequencing techniques (Pfisterer et al. 2021). Once this is accomplished, i.e., as soon as a complete and accurate characterization is available of both the genome and transcriptome of several individual cells within the same neoplasm, the path is open to an improved understanding of tumor heterogeneity as well as the processes of cancer initiation and evolution.

Another area of research in which new technologies promise to help shed light on a fundamental aspect of chromosome behavior in cancer concerns the elucidation of the three-dimensional (3-D) configuration of interphase nuclei. After all, it is at this stage that a cell does most of its actual work, and it seems reasonable to assume that many of the rearrangements that unleash neoplastic behavior occur while the genetic material is decondensed into more than 2 m long chromatin fibers packed within a less than 10 μm diameter nucleus.

Although the possibility, indeed scientific necessity, of moving from mitotic to interphase cytogenetics comes across as thoroughly modern and fresh, it is sobering to realize that some kind of territorial organization of interphase chromosomes within animal cell nuclei was envisaged as early as in 1885, by Carl Rabl. In 1909, Theodor Boveri introduced the term "chromosome territory" in his seminal studies of blastomere stages of the horse roundworm *Ascaris*, arguing that each chromosome visible in mitosis retains its individuality during interphase when it occupies a distinct part of the nuclear space (Figure 17.1). This remarkably prescient concept somehow fell into oblivion, however, and was even considered to have been

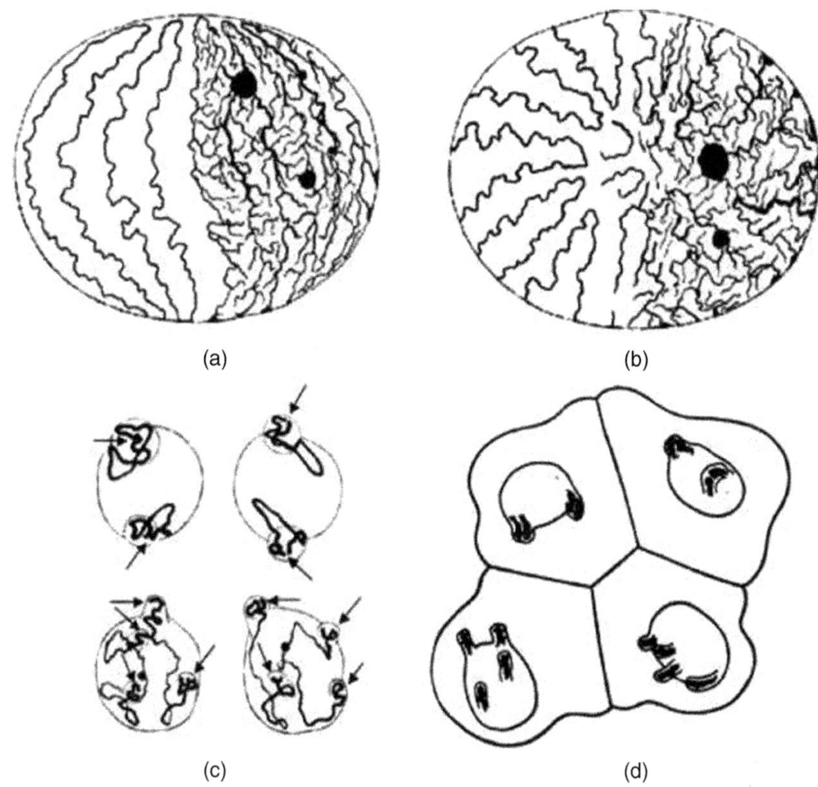

FIGURE 17.1 The past, present, and future meet each other in what is likely to be the new frontier of cancer cytogenetics: The study of neoplastic interphase nuclei. These old drawings, adapted from Cremer and Cremer 2010, illustrate early concepts as to how chromosomes in interphase nuclei may be organized in distinct territories. (a,b) Carl Rabl's hypothetical view (1885) of territorial chromosome arrangement in the interphase nucleus based on his studies of epithelial cells in *Salamandra*. Supposed chromosome territories are built up from primary chromatin threads (a, left side), from which secondary and tertiary threads branch out and form a chromatin network (a, right side). Spindle attachment sites, now known as centromeres, gather at one site of the nucleus, whereas the telomeres cluster at the opposite site (b). (c,d) Drawings by Theodor Boveri (1909) from two *Ascaris* embryos fixed during prophase. Boveri was able to distinguish chromosome ends sticking out in protrusions (arrows in c) of prophase nuclei. He used these protrusions as markers for the nuclear position of the asserted chromosome territories in interphase nuclei. In fixed two-cell embryos, he noted that their nuclear topography was strikingly similar during interphase and prophase, when chromosomes became visible as distinct entities. In four-cell embryos, on the other hand, he typically observed two pairs of nuclei, each carrying distinctly different protrusion patterns (d).

experimentally disproved (for a comprehensive review, see Cremer and Cremer 2010). Instead, the general opinion expressed in textbooks up to the 1980s was that cell nuclei had little internal structure; they were by and large seen as mere containers filled with randomness, like cans of spaghetti (not such a far-fetched analogy given the thread-like appearance of the carriers of genetic material) or even soup. This changed due to pioneering studies performed in the 1970s and 1980s, in particular by Thomas Cremer's group (e.g., Zorn et al. 1976; Cremer et al. 1982) providing partial confirmation of Boveri's allegedly outdated model. Similar results were soon also presented by others and there is now overwhelming evidence from 3-D nuclear architecture studies (Cremer and Cremer 2010; Bickmore 2013; Fritz et al. 2019) that, within this container delimited by a nuclear membrane, chromosomes are not randomly distributed but occupy distinct territories. Inside the nucleus, structural order prevails, with separate domains or compartments corresponding to functional differences. Extremely important in our given context is that large-scale gene regulation is likely to depend on such factors, and yet we know very little about how this type of transcription control is obtained.

It would be surprising if future cytogenetic analyses, be they of normal or neoplastically transformed cells, did not include attempts to unravel the dynamics of this nuclear compartmentalization. What exactly is the nature of the intranuclear structures of importance, how stable/dynamic are they, and how do specific alterations of them correlate with/lead to the altered gene function that plays such a key role in many neoplasms? Answers to these fundamental questions are not available yet.

Another cancer-relevant question that immediately comes to mind when pondering the nucleic structure/function enigma is how and why some neoplasia-specific rearrangements, translocations in particular, are so much more common than others. Of course, there has to be an element of chance and selection, even selection bias, involved – we see only those changes that confer an evolutionary advantage on the cells carrying them – but the suspicion lingers that there may be more to the story than chance plus selection alone. Physical proximity between recombining genes is likely to be one important factor. Some cytogenetic studies of interphase cells have indeed detected a strong correlation between spatial proximity of chromosomes or genes and their translocation frequencies, suggesting that intranucleic architecture does play a role in the genesis and specificity of structural neoplasia-associated chromosome rearrangements (Roukos et al. 2013; Fritz et al. 2019). However, only a very limited number of investigations on selected neoplasms have so far identified preferred chromosome-to-chromosome positional interactions, and the reported proximity patterns seem to be of a probabilistic nature. On the other hand, perhaps this is exactly what one would expect given the predominantly quantitative differences in aberration frequencies seen during cytogenetic analysis of various leukemia and tumor types.

Although it certainly will be exciting to learn more in the years to come about which intranucleic structures or changes are associated with which chromosomal rearrangements in neoplastic cells, causal connections are not going to be easy to prove. First, specific chromosome territories seem to occupy different positions in different cell types and during different differentiation stages. Second, these territories move and intermingle with one another in a manner that is hard, if at all possible, to predict. And third, the perhaps most disturbing point, chromatin fibers sometimes loop out as protrusions from their core territory, allowing transcribing genes, from the same or different chromosomes, to gather in transcription factories at remote sites within the nucleus (Cremer and Cremer 2010, 2019; Fritz et al. 2019). Clearly, proof that spatial proximity among loci in interphase cells determines patterns of chromosomal rearrangements and translocations in cancer will not be easy to establish. Studies to this effect will require concerted efforts involving technological improvements that bring together advanced microscopic methods, subcellular anatomical and biochemical, including molecular genetic, expertise, an intimate knowledge of cytogenetics, proficiency in FISH techniques, and, perhaps most important, computation power beyond anything that is normally relied on in microscopic analyses. The very recent correct prediction by an artificial intelligence network of the 3-D structure of a protein based on knowledge of its amino acid sequence (Callaway 2020) – one of the classic challenges in molecular biology – hints that at least the computational abilities for such endeavors may already exist.

It is easy for technological optimists, such as us, to get completely lost in fanciful speculations about this future that we, alas, are not going to experience. We are very conscious of this danger and want to avoid it; indeed, at the beginning of this chapter the risks inherent in such speculations were mentioned. To make accurate predictions about the future, be they optimistic or pessimistic, smells of hubris, and hubris is the worst human sin. The ancients knew this full well and we agree with them.

It is therefore reasonable to cut our history of the future short at this point; knowledge of the past and present is obviously much more extensive than what we can contribute about things to come. To honor a wise man whose *Tractatus Logico-Philosophicus* was published 100 years ago: Whereof one cannot speak, thereof one must remain silent.

REFERENCES AND FURTHER READING

Bickmore, W.A. (2013). The spatial organization of the human genome. *Annu. Rev. Genomics Hum. Genet.* 14: 67–84.

Boveri, T. (1909). Die Blastomerenkerne von Ascaris megalocephala und die Theorie der Chromosomenindividualität. *Arch. Zellforsch.* 3: 181–268.

Callaway, E. (2020). It will change everything': DeepMind's AI makes gigantic leap in solving protein structures. *Nature* 588: 203–204.

Cremer, T. and Cremer, M. (2010). Chromosome territories. *Cold Spring Harb. Perspect. Biol.* 2 (3): a003889.

Cremer, T. and Cremer, M. (2019). Nuclear compartmentalization, dynamics, and function of regulatory DNA sequences. *Genes Chromosomes Cancer* 58: 427–436.

Cremer, T., Cremer, C., Baumann, H. et al. (1982). Rabl's model of the interphase chromosome arrangement tested in Chinese hamster cells by premature chromosome condensation and laser-UV-microbeam experiments. *Hum. Genet.* 60: 46–56.

Dey, S.S., Kester, L., Spanjaard, B. et al. (2015). Integrated genome and transcriptome sequencing of the same cell. *Nat. Biotechnol.* 33: 285–289.

Fritz, A.J., Sehgal, N., Pliss, A. et al. (2019). Chromosome territories and the global regulation of the genome. *Genes Chromosomes Cancer* 58: 407–426.

Giefing, M. and Siebert, R. (2019). FISH and FICTION in lymphoma research. *Methods Mol. Biol.* 1956: 249–267.

Grimwade, L.F., Fuller, K.A., and Erber, W.N. (2017). Applications of imaging flow cytometry in the diagnostic assessment of acute leukaemia. *Methods* 112: 39–45.

Knuutila, S., Nylund, S.J., Wessman, M., and Larramendy, M.L. (1994). Analysis of genotype and phenotype on the same interphase or mitotic cell. A manual of MAC (morphology antibody chromosomes) methodology. *Cancer Genet. Cytogenet.* 72: 1–15.

Martín-Subero, J.I., Chudoba, I., Harder, L. et al. (2002). Multicolor-FICTION: expanding the possibilities of combined morphologic, immunophenotypic, and genetic single cell analyses. *Am. J. Pathol.* 161: 413–420.

Pfisterer, U., Bräunig, J., Brattås, P. et al. (2021). Single-cell sequencing in translational cancer research and challenges to meet clinical diagnostic needs. *Genes Chromosomes Cancer* 60: 504–524.

Rabl, C. (1885). Über Zelltheilung. *Morph. Jb.* 10: 214–330.

Roukos, V., Burman, B., and Misteli, T. (2013). The cellular etiology of chromosome translocations. *Curr. Opin. Cell Biol.* 25: 357–364.

Suvà, M.L. and Tirosh, I. (2019). Single-cell RNA sequencing in cancer: lessons learned and emerging challenges. *Mol. Cell* 75: 7–12.

Teerenhovi, L., Knuutila, S., Ekblom, M. et al. (1984). A method for simultaneous study of the karyotype, morphology, and immunologic phenotype of mitotic cells in hematologic malignancies. *Blood* 64: 1116–1122.

Zhou, Z., Xu, B., Minn, A., and Zhang, N.R. (2020). DENDRO: genetic heterogeneity profiling and subclone detection by single-cell RNA sequencing. *Genome Biol.* 21: 10.

Zorn, C., Cremer, T., Cremer, C., and Zimmer, J. (1976). Laser UV microirradiation of interphase nuclei and post-treatment with caffeine. A new approach to establish the arrangement of interphase chromosomes. *Hum. Genet.* 35: 83–89.

Afterthoughts

No research, at least certainly none within the realm of medicine, exists in a vacuum. The choice of objects or topics to investigate as well as the methodology involved almost always reflect the zeitgeist, the spirit of the times – these points are obvious, even trivial – but also the organization of institutions where the research takes place is important and revealing. Finally, the prevalent ethical mood and the judicial framework within which one works may facilitate scientific efforts or do the opposite.

A tendency for several decades now has been to separate medical research from clinical medicine, allegedly in order to improve both branches by some organizational magic that is beyond us. In our opinion, and we have worked for many decades as both university professors and hospital doctors, this policy has had more disadvantages than advantages. Of course, we understand that it is possible to argue for some benefits emanating from a clearer division between the two – the strong points of both are strengthened when researchers get more time for research and clinicians more time for patients – but experience has taught us that mostly both suffer more than they gain.

This widening gap between the clinic and the laboratory, continuously and consciously generated by the political establishments in all countries we know, invites several metaphorical associations. C.P. Snow's "Two Cultures" lecture from the late 1950s, in which he lamented the lack of any understanding, even meaningful communication, between the worlds of science and the humanities, comes to mind, as does Rudyard Kipling's even more famous poetic sigh of sadness that "East is East and West is West and never the twain shall meet." Some things *are* inherently incommensurable, to argue otherwise would be plain stupid, but others are not. Snow's original focus was on the shortcomings of the British educational system in preparing young people for their entry into an increasingly science-based world, but we know that this insufficiency now certainly is not restricted to the UK. Especially since the curricula of most medical schools currently seem to be geared more toward acquiring "communication skills" and other abilities typical of "soft" disciplines rather than an as profound as possible

Abnormal Chromosomes: The Past, Present, and Future of Cancer Cytogenetics.
Sverre Heim and Felix Mitelman.

understanding of "hard-core" biology, close interactions between practicing clinicians and those who work in the laboratory are difficult to establish and maintain. Beyond any shadow of doubt, it is to the detriment of both worlds, clinical medicine and research, that so few MDs nowadays acquire serious lab experience. Translational medicine cannot work well without competent translators; we should not fool ourselves by believing otherwise.

The need for hospital medicine and laboratory investigation to work hand in hand is particularly obvious in cancer cytogenetics. Whether genomic investigations of neoplastic cells are performed for clinical or research purposes matters little or not at all; an interesting, acquired chromosome aberration is one of nature's riddles regardless, one that begs to be examined further. The analyses are often the same in both situations, only our thinking around the case differs depending on the context.

These reflections bring us to comment on another facet of modern – or should we say postmodern? – scientific medicine, one that again sadly illustrates how the best in many instances has developed into an enemy of the good. We have in mind the increasingly bureaucratized ethical requirements that in practice now prevent many of the more low-key investigative efforts that, in spite of what we wrote in the paragraph above, continue to play a major role in our branch of laboratory research. There is so much paperwork involved if one wants to utilize diagnostic analyses for research purposes that many colleagues refrain from even attempting to do so. To those who opine that this must surely be an exaggeration, we can only state that we know from much too long practice that the extensive red tape certainly does not feel like "simple procedures to safeguard patient integrity and obtain permission or informed consent," not when you have to find your way through them. Good, highly ethical colleagues have said the same, often more forcefully. Likewise, most patients are at a loss to understand why they have to answer consent-seeking letters before hospital doctors they have never even seen can do what they deem best with the remains of neoplastic samples removed for therapeutic or diagnostic purposes. They nearly always agree to participate, saying that they are glad to support science as much as they can in spite of knowing full well that they themselves are not going to be the beneficiaries of whatever new knowledge is gained, but often they, too, find the motions useless. When push comes to shove and death becomes a threatening reality, contemporary patients nearly always choose to trust their doctors. Too much bureaucracy – and now there *is* too much of it surrounding research that ideally should run smoothly and concomitantly with diagnostic efforts, at least in our field of medicine – is a hindrance preventing the easy scientific use of diagnostic data. In short, many of the existing ethical safeguards are no longer working in the patients' best interest.

The major ethical problem today is, beyond any shadow of doubt, that *too little* research is done to elucidate what happened in the many "Devil's experiments" that come our way – we are talking about cancer cases – which in turn leads to clinical progress being belated, not that patients are put at risk by the acts of uncaring doctors.

An unwritten and invisible keyword in all the deliberations above is *quality*. Much lip service is nowadays being paid to exactly that in health institutions, university hospitals included, stressing the need to "put quality first" or "ensure quality control," but most of it rings hollow. In practice, it is not actual quality that is being monitored, controlled or even encouraged, but rather standardized administrative procedures for filing, registration, answering and the like. This becomes yet another bureaucratic burden on those who do the actual work, and hence a negative incentive to walk the extra mile toward true excellence.

What is the immediate product of cytogenetic investigations? This probably sounds like a trick question and to some extent it is, but the answer we are after is: *Data*. Microscopic examinations of well-spread and properly banded dividing chromosomes generate innumerable bits of sensory data that are processed and sieved, automatically as well as consciously, into a much smaller number of relevant pieces of information that can be used, together with the cytogeneticist's existing knowledge, to generate a reliable picture of the neoplastic cells' genomic constitution, their karyotype. Since this is visual information and the retina is an integral part of the brain, the mostly subconscious or automated nature of this process is, we think, fairly easy to account for: The brain does brain stuff with considerable ease regardless of which parts of it are interacting. This, too, is part of the advantage of working with information obtained by seeing.

The next step in the epistemological chain or ladder that forms the backbone of all investigative procedures is to translate the data into *knowledge*. What does it mean, that which we saw? This relies on a much more conscious effort as well as medical, genetic, and scientific experience coupled with an ability to think reasonably straight. It is not easy, but it can certainly be learned by many.

The third step is to synthesize the newfound knowledge with what we already possess of preexisting knowledge and general competence both within the cancer cytogenetics field and generally into an *understanding* of what has gone wrong in the transformed cells. How do the data at chromosome and cell level collate with both the patient's best interests and our understanding of tumorigenic processes in general? This question includes, whenever possible, reflections on or ideas about how the new cytogenetic information can possibly be translated into novel therapeutic strategies. Again, knowledge at this level, such deep understanding of the subject matter, is difficult to obtain but it is a most worthy goal to strive for.

Finally, on rare and precious occasions, a feeling of newfound *wisdom* may be granted us at the very top of the epistemological ladder. Such moments should certainly be savored although they are few and far between. They belong to the things of which "one cannot speak," a circumstance that by no means should be interpreted as a negation of their (however infrequent) existence.

The metaphorical ladder above is certainly a two-way structure, in the sense that a degree of correspondence, or perhaps inverse correspondence, exists between the experimental science within cancer cytogenetics and our thinking about what we do. At one moment, the cytogeneticist descends looking for smaller structures at higher resolution levels, at another the effort is mainly synthetic, attempting to find the meaning of whatever abnormal structures have been seen. This is how the epistemological ladder is supposed to function theoretically, as already outlined, but surprisingly often the same description holds for how cancer cytogenetics is performed in practice, too. Not rare are the moments when, during analysis, you stop and sigh, wishing that you could gain access to at least one or two really high-quality metaphase spreads or chromosomes that would help you get to the bottom of things and conclude the work satisfactorily. If you find them, then that may well be all that is needed for the diagnostic problem to be solved, for as all experienced cytogeneticists know: Chromosomes do not lie.

"All" we have to do, then, is learn how to "speak chromosome." But that is much too difficult, say some. Geneticists, doctors, and technicians of today cannot be expected to learn such things, to put in all the hours that make for the "green fingers" necessary to do good lab work, nor can they spend the necessary time to hone the analytical and synthetic skills needed afterwards in order to see what they actually saw. Well, cancer cytogenetics may not be easy, we readily grant you that, but the same goes for playing chess, tennis or the violin. If it were easy – we are back to cancer chromosomes now, not our unsuccessful careers in the other fields mentioned – probably more people would have done it and the "genomic mine" full of hidden nuggets would by now have been emptied. Many worthy activities in life are hard to learn. A science that relies on things being easy does not command much respect, nor does it come across as realistic.

Instead, the said gold mine probably still retains most of its treasures; we are far from knowing all there is to be known about the chromosomal changes that govern neoplastic transformation. To those few who have the time, will, curiosity, and intellectual stamina: Keep digging, your efforts will not be in vain! Be certain that as you gradually learn to speak to the cells and chromosomes, they will answer you, and the exchange will not be void of meaning.

Index

Abnormal Chromosomes: The Past, Present, and Future of Cancer Cytogenetics.
Sverre Heim and Felix Mitelman.